TEACHING AND COLLECTING
TECHNICAL STANDARDS

PURDUE INFORMATION LITERACY HANDBOOKS

The Purdue Information Literacy Handbooks series publishes works that present and discuss in-depth practices, research, and theory that advance information literacy. A key resource for academic librarians, researchers, educators, and students, PILH pushes the boundaries of the field by exploring information literacy in social, educational, and workplace settings while deploying innovative methods that investigate current and emerging ideas. Global in scope and written by practitioners, books in this series are meant to combine theory and practice. Topics of interest for the series include but are not limited to teaching and learning; human rights and social justice; disciplinary or professional communities; specialized literacies, such as data, digital, and archival; and media, democracy, and civic discourse.

SERIES EDITOR

Clarence D. Maybee, Purdue University

SERIES EDITORIAL ADVISOR

Christine Bruce, James Cook University

OTHER TITLES IN THIS SERIES

Teaching Information Literacy and Writing Studies: Volume 2, Upper-Level and Graduate Courses
Grace Veach (Ed.)

Teaching Information Literacy and Writing Studies: Volume 1, First-Year Composition Courses
Grace Veach (Ed.)

Data Information Literacy: Librarians, Data, and the Education of a New Generation of Researchers
Jake Carlson and Lisa R. Johnston (Eds.)

Integrating Information into the Engineering Design Process
Michael Fosmire and David Radcliffe (Eds.)

TEACHING AND COLLECTING TECHNICAL STANDARDS

A Handbook for
Librarians and Educators

edited by
Chelsea Leachman, Erin M. Rowley,
Margaret Phillips, and Daniela Solomon

Purdue University Press • West Lafayette, Indiana

978-1-61249-820-1 (hardback)
978-1-61249-821-8 (paperback)
978-1-61249-822-5 (epub)
978-1-61249-823-2 (epdf)

Cover image: Layout by Purdue University Press using the following assets:
monkeybusinessimages/iStock/Getty Images; gorodenkoff/iStock/Getty Images;
g-stockstudio/iStock/Getty Images

Contents

PART IV. CASE STUDIES

Preface

GOALS FOR THE BOOK

Defining "standards" is a difficult task. Standards, in terms of this book, do not tend to have a "standard" definition, as it were. Experts from various areas often take many sentences or even paragraphs to define and describe what standards are and their importance. A broad definition from Sullivan's 1983 book states the standards "are a category of documents who function is to control some aspect of human endeavor" [1]. As Sullivan admits, "it is an exceedingly large field." A section of Crawford's 1985 book was devoted to defining technical standards where it is stated that "technical standards are definitions or specifications; they communicate agreement on sharing techniques" [2]. The section goes on to provide additional information and context on what, exactly, technical standards are and aim to do.

Standards are an essential source of information for providing guidelines during the design, manufacture, testing, and use of whole products, materials, and components [3]. To prepare students for the workforce, universities are increasing the use of standards within the curriculum. Engineering employers believe it is important for recent university graduates to be familiar with standards [4–6]. One way for students to become independent and highly competent at finding standards information is through integration into the curriculum. Despite the critical role standards play within academia and the workforce, little information is available on the development of "standards information literacy." Standards information literacy includes the ability to understand the standardization process; identify types of standards; ability to identify standards, locate, evaluate, and use standards effectively.

The information regarding standards most commonly is provided by individual standards developing organizations (SDOs). While the information from an individual SDO is helpful to specific fields, the literature aimed at students, librarians, and course instructors related to standards is either decades old or is limited in scope. Therefore, the need for an up-to-date and comprehensive resource on standards history and development, as well as how standards can be integrated into information literacy instruction was needed.

Standards information literacy is many times co-taught by librarians and engineering course instructors. Libraries and librarians are a critical part of standards education and much of the discussion has been focused on the collection of and access to standards within libraries. However, librarians also have substantial experience in developing and teaching standards information literacy curricula. With the need for universities to develop a workforce that is well-educated on the use of standards, librarians and course instructors can apply their experiences in information literacy towards teaching students the knowledge and skills regarding standards that they will need to be successful in their field.

This book captures the experience of librarians and course instructors on the use of standards within the academic practice in higher education. To meet the academic and workforce needs, the goals of this book are to:

- Highlight the history of standards
- Explain the standardization process and types of standards
- Establish the value of standards education within the academic curriculum
- Demonstrate standards information literacy in academic practice
- Demonstrate standards collection development in academia

As a unified presentation of standards information for both instructors and librarians, this book illustrates a comprehensive model for institutions to use when building a standards information literacy curriculum.

A primary use of this book is to serve as a resource for engineering librarians and engineering educators to use and modify the standards-in-practice lessons as needed for their local context. We believe these lessons will be particularly useful for first-year engineering courses, engineering design courses in all engineering disciplines (all years, including senior capstone), engineering management courses, technical communications courses for engineering and technology students, materials and testing courses, and metrology courses. However, as standards are used in other disciplines outside of engineering, this book would also be useful to subject librarians and educators working in other areas such as business, health sciences, and law. The book contains chapters and case studies specifically aimed at these other disciplines and it is the hope that this book is used as a resource by all academic areas impacted by standards. Lastly, the book could be a helpful supplemental resource for library school courses focused on engineering and technology resources, or again, resources related to business, health science, and law.

FRAMEWORK FOR THE BOOK

This book is organized into four parts: a standards overview, standards access and collection development to support information literacy, standards curriculum integration and requirements, and case studies using standards that can be used in a variety of class settings from undergraduate to graduate level. While our intent was to cover how standards are integrated into curricula, as well as the importance of standards information literacy, we quickly realized that this book may have multiple audiences. Therefore, introductory information on standards including a history of standards, types of standards, and the standards development process is covered for the standards novice, regardless of profession. As current and future librarians may also find this book useful, we expanded our scope to also include standards collection development information.

It is important to note that the world of standards continues to evolve over time, and as such, information changes. We made every effort to consult a wide range of sources when writing this book; however, the impact of standards is far-reaching. Additional reading and resources are highlighted throughout the book for further or more in-depth information on select topics. Links to supplementary resources and readings are provided wherever possible; at the time of publication, all links were live, but bear in mind how quickly some URLs can change.

Ultimately, we are hopeful that this book proves to be a useful resource for anyone interested in learning more about standards or incorporating standards into their higher education curriculum.

REFERENCES

1   C. D. Sullivan, *Standards and standardization: Basic principles and applications.* New York: M. Dekker, 1983.

2   W. Crawford, *Technical standards: An introduction for librarians.* White Plains, NY: Knowledge Industry Publications, 1986.

3   M. Phillips and P. McPherson, "Using Everyday Objects to Engage Students in Standards Education," in IEEE Frontiers in Education Conference (FIE), Erie, PA, 2016: IEEE, https://doi.org/10.1109/FIE.2016.7757698.

4   B. Harding and P. Mcpherson, "What Do Employers Want in Terms of Employee Knowledge of Technical Standards and the Process of Standardization," in *2010 ASEE Annual Conference & Exposition*, Louisville, Kentucky, June 20 2010, https://doi.org/10.18260/1-2--16474. Available: https://peer.asee.org/16474.

5   Jeffryes and M. Lafferty, "Gauging workplace readiness: Assessing the information needs of engineering co-op students," *Issues in Science and Technology Librarianship*, vol. 69, no. 69, 2012, https://doi.org/10.5062/F4X34VDR

6   N. Waters, E. Kasuto, and F. McNaughton, "Partnership between Engineering Libraries: Identifying Information Literacy Skills for a Successful Transition from Student to Professional," *Science & Technology Libraries*, vol. 31, no. 1, pp. 124–132, 2012, https://doi.org/10.1080/0194262X.2012.648104.

PART I
Standards Overview

1

Introduction to Standards

Chelsea Leachman, Washington State University

Whether people are aware or not, standards affect our everyday lives, from the transportation we use, to the light bulbs used within homes and buildings, to the barcodes used to purchase items at the store. The simplest definition of *standards* is to gain a level of quality or attainment of an idea or thing, including items used as a measure, norm, or model in comparative evaluations. Historically, standards have been established in engineering, science, technology, health care, business, and many more disciplines. Standards are used to norm criteria, methods, processes, and practices. Standards and standardization are often developed in response to new knowledge and understanding of products or processes. While current standards developing organizations (SDOs) react to advances in the 21st century by creating or changing existing standards, standards date back to ancient civilizations to advance commerce in societies by standardizing the payments for goods and services. Standardization continues to foster trade around the globe by creating standardized processes at all levels, from local to international.

BRIEF HISTORY OF STANDARDS

The history of standardization can be told in many different ways and lengths. In this volume, the authors wanted to include a brief history of standardization to give the reader a foundation for how standards and standardization have shaped the world and will continue to impact everyday lives. When focusing on the history of standards and standardization, the reader should remember that the development of standards and standardization is ongoing [1]. From the beginning, the full impact of standardization on the world cannot be predicted. Even in the development of standards and standardization today, the full effect of one standard or standardization is hard to realize at the time of the standard's creation. It isn't until much later that the total impact can begin to be measured and understood.

From the earliest civilizations, standardization, while often not formally documented, has impacted humanity through rituals or ceremonies [1]. The earliest standards fall into four categories: counting, shape, weight, and time [1, 2, 3]. These four categories are universal and express themselves differently depending on the culture. When looking for other early examples of standardization throughout our culture, spoken language and the advent of written language brought about standards for the visual expression of written language. With alphabetic and character-based written languages, the need for standardization became apparent with the ability either to spell the same word in different ways or to use slightly different characters for the same word [1]. While there are dialectic differences worldwide, there can be a common understanding of the meaning and a reduction of errors with standardization.

With the creation of monetary exchange, there became a need for formalized standardization [1]. The earliest documentation of standardization comes from economic exchanges recorded in ancient civilizations' texts, where there are records of standardization for buildings, marriages, and crimes [1]. Standardizing these was to ensure fairness in commerce and codes [1].

During the 17th and 18th centuries, the Age of Enlightenment, standardization gained popularity with the advancement of scientific methods and technological developments [1]. During this time of enlightenment, scientists debated and discussed the true nature of experimentation and the resulting outcomes. Out of these debates and discussions came four rules of the scientific method from René Descartes in 1637, who stated that scientists (1) were not to accept anything for true that they did not know to be such; (2) to divide the scientific question into as many parts as possible to examine; (3) to conduct the experiment in a logical, step-by-step order; and (4) to make enumerations so complete and review so general that we can be assured nothing was omitted [1]. Many of these rules still apply to science conducted today and help scientists worldwide understand the process in which an experiment or problem has been examined. In addition, many experiments have standardized protocols that scientists must follow to ensure understanding throughout the scientific community. The standardization of the scientific process is similar to the languages discussed above in that scientific experimentation and study can be described as a form of language.

Another example of standardization throughout history was the need for consistent timekeeping. The standardization of timekeeping was created out of the need for consistency among societies and communities. Whether it be keeping track of the season or a precise time of day, the development of standardized timekeeping changed and regulated commerce activities worldwide. One early example of the need for timekeeping was the United States railroad system in the late 19th century, which "had not fewer than fifty-three different standards" to set their clocks by [1]. With so many different standards to set time, there was a real fear that trains could collide with one another when crossing into a different time area, or what is referred to today as a *time zone*. Timekeeping was solved through "railroad time," which created standard time zones for all clocks to be calibrated [1].

Institutions that became highly interested in standardization were military systems. Militaries needed to be well organized, and uniformity was imperative. Standardization within the military is most frequently related to the early need for weapons standardization. When standardizing weapons, one of the tasks was to address technical problems by having interchangeable parts. One example of a consensus standard within the military came about during the late 19th century. With the advent of mass production and the interchangeability of parts, the concern for modern standardization became the concern of engineers and industrialists. The experience of Eli Whitney [1], a gunmaker, is often cited as an example:

> In 1798, our [USA] government was in need of more and more arms. Jefferson, then Vice President, signed a contract which bound Eli Whitney to supply ten thousand muskets in two years. At the end of the first year, only five hundred had been delivered, production of less than two a day. The two years expired and so did Whitney's contract. Necessity became the mother of invention. Urged by the government, Whitney submitted to a board of experts the

assembly parts of ten muskets and in their presence assembled from ten identical barrels, ten identical stocks, and ten identical triggers, the first ten standardized rifles. By introducing the principle of interchangeable parts for armament production, he thus became the father of mass production for war purposes.

Outside of the standardization of military weapons was the need for other military supplies to be consistent over longer periods of time from different manufacturers. One of these supplies was food that could last and be carried long distances. Out of this need, the process of canning in the food-processing industry was created by Nicolas Appert in 1809 with a bounty paid by Napoleon to whoever could create an inexpensive and effective way of preserving fresh food [1].

Standardization not only was gaining popularity in factories for the creation of products. During the Industrial Revolution, the practice of medicine also was undergoing a change in the education of medical professionals and medical research. Important medical standardization ranges from the formalization of patients' files or records kept by physicians and hospitals to the development of medical equipment [1]. The changes in the early 20th century to patient medical records were endorsed by the American College of Surgeons through a hospital standardization program and allowed medical data to be compared throughout medical systems. With the advancement of medical equipment, such as X-ray machines, scanners, and other medical diagnostic devices, standardization has allowed practitioners to analyze and understand the results of medical tests regardless of familiarity with the specific equipment [1]. Finally, one of the big advances of standardization in medicine was the *International Classification of Diseases* (ICD-10), which allows medical practitioners a standard diagnostic classification [4].

The law is another profession that has adopted standards and standardization processes. As with the professions mentioned above, the legal system as a profession has created its own set of professional standards through the American Bar Association. The first set of standards adopted by the American Bar Association was in 1908 with the Canons of Professional Ethics, the first national standard of ethics for lawyers [5]. The first professional standards were created in 1964 and are used by judges, prosecutors, attorneys, legislatures, and scholars in the criminal justice system. Along with professional standards, education standards for law students were adopted in 1921 by the American Bar Association as the Standards and Rules of Procedure for Approval of Law Schools [5]. The adoption of standards for education led to the accreditation of law schools beginning in 1952 [5].

Agriculture has also been subject to standardization by focusing on production and productivity. This standardization comes in many forms, from crop varieties to seeds, fertilizers, farm equipment, and food safety. Due to production and economic needs, farmers were "encouraged to plant a single variety" [1] of crops instead of various strains due to the marketability of a uniform crop. Following the guidance to plant a single crop, standardization followed with uniformity of seeds and fertilizers to ensure productivity when planting. One way fertilizers were standardized was by labeling the product and how to use it. Farm equipment also moved toward standardization to utilize standard fertilizer and seeds to increase productivity using mechanical systems for planting and harvesting. Once the crops were harvested and with the uniformity of crops, the development of packaged products increased. Food safety and food quality assurance became needed around the world. The first food safety laws appeared in the late 19th century and used scientific tests to measure the food itself [1].

While products and processes were some of the first governmental standards, one of the first voluntary standards came later when the United States and Canada needed to address the challenges with the international railroad system. During the late 19th century, the United States and Canada needed to address the challenge of nine different railroad gauges. The differences in railroad gauge— the distance between a pair of rails—was causing issues with interregional travel [6]. The gauge difference came from the ability of individual private companies to choose their track gauges with minimal government regulation at the time. Specific gauge preferences were only affected where new rail lines were being constructed, and there was an interest in compatibility with existing neighboring lines. As the railway system within the United States continued to grow, the expectation of compatibility increased; however, the choice of rail line gauge for new lines was left up to the chief engineer, who commonly used gauge lines previously utilized. By the 1860s, the railway system began to see an increase in interregional transportation, and companies sought to maximize the value of the railways. Due to the increasing demand, the railway gauge differences were resolved, and 4 feet 8.5 inches (1.4 meters) became the standard gauge for the railway network within the United States and Canada, making interregional travel and commerce possible [6].

Moving into the Industrial Revolution, fueled by innovation, science, and technology, standardization was needed for increased production and safety of products; there also came the need to protect factory workers. The awareness for factory worker safety came from the well-known boiler explosion in 1905 at the

Grover Shoe Factory in Brockton, Massachusetts, which tragically injured 150 and killed 58 people [7]. The chief engineer, David B. Rockwell, used the older boiler built in 1891 as the new boiler underwent maintenance. The older boiler had been maintained and recently inspected. The boiler tragedy was due to a crack that formed behind one of the lap joints, "two pieces of steel that overlap for several inches and are held together with steel rivets that inspectors could not see" [7]. The American Society of Mechanical Engineers (ASME) created the "Boiler and Pressure Vessel Code" in response to this tragic event.

HISTORY OF STANDARDS DEVELOPING ORGANIZATIONS

Similar to the history of standards and standardization, this title will not cover all individual standards developing organizations, but will focus on the organizational history of governing and large organizations. As standardization moved forward, the need for individual interest groups to create their own standards grew. Out of these groups' interests, there came a need to coordinate standards development and have a national consensus for standardization. Late in the 19th century, scientific communities worldwide started to gather to agree upon terminology and measurements. The first organization to be formally created in the United States by Congress in 1901 was the National Bureau of Standards (NBS), which would later become the National Institute of Standards and Technology (NIST) in 1988 [8]. The National Bureau of Standards was created as the authority on domestic measurements and standards laboratory in the United States. One of the major issues at the turn of the 20th century was the inconsistency of measurements; for example, there were at least eight different measurements for gallons and four different measurements for feed being used. The National Bureau of Standards conveyed the first National Conference on Weights and Measures in 1905 to write laws and standards to distribute to inspectors and create a fair marketplace [8]. Other areas that the NBS influenced through standardization are the weights and measures of railroad cars, electric safety codes, radio transmissions, fire safety, radiation safety, photography, explosives, and computers.

In addition to the NBS, the International Metric Commission was created in 1971 and later became the International Bureau of Weights and Measures [9]. The United States signed on to the creation of the group without ever implementing the metric system, which was an ongoing point of contention throughout the scientific community [9]. Scientists and engineers continued to meet at expositions and conferences to discuss common rules and understandings throughout the scientific community.

Out of the global need for scientific consensus and for national consensus in the United States, the following groups joined together in 1918 to collaborate and become the founding members of what is known today as the American National Standards Institute (ANSI): the American Institute of Electrical Engineers, the American Society of Mechanical Engineers, the American Society of Civil Engineers, the American Institute of Mining and Metallurgical Engineers, the American Society for Testing and Materials, and the U.S Departments of War, Navy, and Commerce. When ANSI was initially established, it was called the American Engineering Standards Committee [10]. ANSI oversees standards in the United States. Many of the individual organizations that joined to create ANSI were founded in the late 19th century as the engineering profession sought to raise its status in the industrializing world [10].

Throughout the 1920s, ANSI approved its first standard on pipe threads and started a major project to coordinate national safety codes to prevent accidents [11]. ANSI's first American Standard Safety Code was approved in 1921 and covered the protection of the heads and eyes of industrial workers [11]. National standards developed in the 1920s included mining, electrical and mechanical engineering, construction, and highway traffic.

During World War II, ANSI was prepared with a War Standards Procedure that helped "accelerate the development and approval of new standards needed to increase industrial efficiency for war production" [10]. American war standards were produced with the help of more than 1,000 engineers, and were intended to ensure quality control and safety of military and civilian products. In 1926, ANSI hosted a conference where the International Standards Association (ISA) was created [11]. The ISA would become the International Organization for Standardization (ISO) after World War II when ANSI and national standards bodies from 25 countries joined forces [10]. ISO was created to promote international standards and facilitate the global unification of industrial standards. The International Electrotechnical Commission (IEC) is the sister organization to ISO and other international and regional standardization bodies.

Internationally, consensus standardization was gaining popularity within the science and technology communities, leading to the first standard approved by ISO in 1951 for the reference temperature for industrial length measurements [11]. Since this first standard was approved, it has been updated many times. As technology progressed through the 1950s, standardization bodies helped industry and governments create standards for developing nuclear energy, information technology, and electronics.

ISO published the first International System of Units (SI) in 1960 and set one unit for each quality, for example, the *meter* for distance and the *second* for time. The objective of the SI system is to reach worldwide uniformity in units of measurement. Throughout history, standardization has been essential to safety and the economy worldwide. Standards are being created as new technologies are being developed and changed as technologies are used in different and unique ways.

PURPOSE OF STANDARDS

Standards have been used informally and formally for centuries by people worldwide. Civilizations have developed a consensus in any given field to assure quality, methodology, safety, or operations through these rules. Today's standards are written for the following purposes: safety and reliability, reduction of cost and waste, interchangeability, and societal organization.

Firstly, the safety and reliability of standards and standardization are of great importance for consumer and worker safety. A potent example of safety and reliability is when consumers buy a product such as over-the-counter medicine. How would a consumer know if the different medicine brands have the same quality? The standard for over-the-counter medicine comes from the United States Pharmacopeia (USP), a standard used by medicine manufacturers to ensure that the product meets the national standard for strength, quality, and purity [12]. Without this standard, products on the market could have different strengths or qualities, leading to consumer safety issues.

Secondly, the reduction of costs comes from both the manufacturing side of the development of products and the development of procedures for manufacturing. With common systems in place, people can efficiently and effectively follow a standard approach to either design, manufacture, communicate, or work.

Thirdly, the interchangeability of standards and standardization make mass production possible and set a baseline for products or services to allow innovation [12]. Interchangeability also promotes business by fostering a global economy through the ability to produce and sell products worldwide.

And finally, the organization of societies. Standards and standardization touch our lives in many different ways every day. Large-scale organization of societies would be complex without standardization, from the standards of traffic lights to the ability to understand the justice system. Standards help society function from the smallest bolt to social structures.

WHO USES STANDARDS?

Often, the first time people use and become aware of standards is during a professional training program, apprenticeships, or a time of need; however, as mentioned above, standards are in every sector of human life. To name a few, technical standards reach a broad spectrum of disciplines from business, engineering, and health care. While each discipline's use of standards varies, they are in place to provide safety for consumers, employees, and the environment.

When considering a product development process, the designers engineering a product will seek out existing standards early in the design process to help with decisions. As the product is moved through design and development, the subsequent use of standards aid in product testing. Finally, as the product moves into manufacturing, the employees producing the product might have company-specific standards to follow throughout the manufacturing process. At each step of developing a product, there are standards used by different people for different needs.

While standards are often used explicitly by professionals, everyone is a consumer of standards and standardization in the products and services purchased and used daily.

STAKEHOLDER BENEFITS OF STANDARDS AND STANDARDIZATION

While everyone is a stakeholder in developing and using standards, interest groups have specific benefits. Consumers benefit from the development and use of standards and standardization through the safety and uniformity of products and services. For example, the health care system has medical equipment developed and tested using standards and standardization. For consumers or the public using the health care system, standards and standardization come to light when receiving care from medical professionals. Therefore, it is in the consumer's best interest to have standards behind developed products.

Companies or corporations developing products or services benefit from standards and standardization to be efficient and effective. For example, companies must understand the safety requirements and manufacturing standards when developing new products. Understanding the manufacturing standards and standardization will benefit the company in cost savings during product development and increased production efficiency. Companies and corporations also can have standardized internal processes or procedures that employees need to follow, which will help with the efficiency of the company long term. Some industries, such as car manufacturers or health care systems, have worked together to create standards for that industry, leading the stakeholders to benefit from the collective knowledge.

Other stakeholders that benefit from developing and using standards are government agencies. While government agencies worldwide benefit from standardization, this book focuses on U.S. government agencies, including all levels, from federal to state to municipalities, when referring to government agencies. Government agencies can adopt standards from SDOs or companies where applicable, or they can create their own standards for their specific needs. When government agencies use standards that have already been developed, it saves them the time and energy to create their own; however, if they create a technical standard for a specific need, they also have industry and scientific knowledge to help them create the standard.

SUMMARY

From the beginning of human history, people have found ways to standardize the world around them. Through commerce and daily living systems, people were able to work together to make systems that would bring people together across different cultures to work together. Modern-day standards and standardization didn't come about until the late 19th century and vastly changed many professional landscapes. Today many people learn about standards and standardization throughout their professional training, while many consumers are usually unaware of the amount of standardization around them daily. Once one becomes aware of standards and standardization, it is hard not to notice them. While standards and standardization are meant to change with new technologies and understanding, we must recognize the past and how far standards and standardization have come. Many standards and standardization were developed out of a dire need for safety, both through products and systems, but with an understanding of the past, we can continue to question current standards and standardization practices to push the process further for the betterment of humanity.

REFERENCES

1   L. Busch, "Standardizing the world," in *Standards: Recipe for Reality*. Cambridge, MA: MIT Press, 2011, pp. 82–149.

2   C. D. Sullivan, *Standards and standardization: Basic principles and applications*. New York: M. Dekker, 1983.

3   L. C. Verman, *Standardization, a new discipline*. Archon Books, 1973.

4   World Health Organization, *International statistical classification of diseases and related health problems*, 10th ed.

5   American Bar Association, *Standards archives*. https://www.americanbar.org/groups /legal_education/resources/standards/standards_archives (accessed Oct. 15, 2021).

6   D. J. Puffert. "The standardization of track gauge on North American railways, 1830–1890." *The Journal of Economic History, 60(4)*, 2000, pp. 933–960.

7   New England Historical Society, *The Grover Shoe Factory disaster shakes the nation in 1905*. https://www.newenglandhistoricalsociety.com/grover-shoe-factory -disaster-shakes-nation-1905 (accessed Feb. 19, 2021).

8   National Institute of Standards and Technology, *NIST timeline*. https://www.nist .gov/timeline (accessed Aug. 24, 2022).

9   C. N. Murphy and J. Yates, *Engineering rules*. Baltimore, MD: Johns Hopkins University Press, 2019.

10 American National Standards Institute, *ANSI history*. American National Standards Institute. https://www.ansi.org/about/history (accessed Feb. 19, 2021).

11 International Organization for Standardization, *The ISO story*. https://www.iso .org/the-iso-story.html (accessed Mar. 10, 2021).

12 United States Pharmacopeia, *Reference standards*. https://www.usp.org /reference-standards (accessed Nov. 21, 2021).

2

An Exploration of Types of Standards

Daniela Solomon, Case Western Reserve University

INTRODUCTION

Standards are the foundation for the development and implementation of new technologies, products, processes, and services by promoting interoperability, reliability, safety, and quality of materials. By creating a common global language for product development and safety, standards are critical for a functional global economy.

Standards are used by industries and professionals worldwide and contribute to and enhance many aspects of our daily lives. Their development is demanded by the needs identified by the market.

Standards are usually formal documents that establish uniform specifications and procedures designed to maximize the performance, quality, and safety of products, processes, and services. Standards may be developed through a company, consortia, industry, or the regulatory/government level and have different compliance expectations depending on the developing organization. Standards may have various formats depending on the application, content, geographical coverage, user group, and so forth.

DEFINITION(S)

Standards definitions vary within the various organizations implicated in the development process. Interestingly, there is no "standard" definition for standards. A general definition would be "written agreements containing technical specifications or other precise criteria that may contain rules, guidelines, or definitions of characteristics" with the purpose to ensure the quality, safety, and interoperability of materials, products, processes, and services [1].

The International Organization for Standardization (ISO) and its sister organization, the International Electrotechnical Commission (IEC), define *standards* as,

> a document, established by consensus and approved by a recognized body, that provides, for common and repeated use, rules, guidelines or characteristics for activities or their results, aimed at the achievement of the optimum degree of order in a given context [2].

In the United States, the National Institute of Standards and Technology (NIST) defines standards as documents "developed or adopted by voluntary consensus standards bodies, through the use of a voluntary consensus standards development process" that includes:

> (i) common and repeated use of rules, conditions, guidelines or characteristics for products or related processes and production methods, and related management systems practices; (ii) the definition of terms; classification of components; delineation of procedures; specification of dimensions, materials, performance, designs, or operations; measurement of quality and quantity in describing materials, processes, products, systems, services, or practices; test methods and sampling procedures; formats for information and communication exchange; or descriptions of fit and measurements of size or strength; and (iii) terminology, symbols, packaging, marking or labeling requirements as they apply to a product, process, or production method [3].

Standards should be developed taking into consideration science, technology, and professional expertise with the goal of ensuring community benefits.

DE FACTO AND DE JURE STANDARDS

Generally adopted and market-dominant standards are known as *de facto* standards (e.g., the QWERTY keyboard). Standards that have been developed and approved by formal authorities are known as de jure standards.

De jure standards are developed by committees consisting of experts in the field. One very important aspect of de jure standards development is the requirement to reach consensus before a standard is adopted, where consensus is defined as "general agreement, characterized by the absence of sustained opposition to substantial issues by any important part of the concerned interests and by a process that involves seeking to take into account the views of all parties concerned and to reconcile any conflicting arguments" [2]. Consensus can be reached without reaching unanimity.

When developed for a company's internal use or for the consortia members, standards have limited consensus, and their applicability is limited to the developing constituents; however, when standards are developed for wider adoption, reaching a general consensus among all the development process participants is required. For the purposes of this chapter, we are going to look at the classification of de jure standards.

MANDATORY/VOLUNTARY STANDARDS

The standardization process is specific to each country. Most countries have a designated organization as the major standards developer that has strong relationships with the local government.

In the United States, the standardization system includes both governmental and nongovernmental organizations. The nongovernmental standards developing organizations (SDOs) are subject to the American National Standards Institute (ANSI) accreditation and are required to follow the guidelines imposed by ANSI. The SDOs ensure wide representation in the development process by bringing together all interested stakeholders—manufacturers, consumers, representatives of government, and academia.

One unique aspect of the standards developed by SDOs is that they are *voluntary*, meaning both that participation in the development process is voluntary and that compliance with the resulting standards is voluntary. Participation of federal representatives in the development of these standards is encouraged to ensure that the standards will meet both public and private sector needs [4].

In the United States, standards are *mandatory* when they are set or adopted by the government and can be either procurement or regulatory standards. Additionally, when government regulations refer to privately developed standards, those standards also become federal, state, or local law. This is known as *incorporation by reference*. OMB A-119 is a federal policy document that encourages all federal agencies to use private-sector standards versus government-unique standards. Voluntary standards also become mandatory if they are part of a business contract or when a product is marketed as fulfilling the requirements of voluntary standards.

LOCAL, REGIONAL, NATIONAL, AND INTERNATIONAL STANDARDS

Standards can be adopted at various geographical levels: local, regional, national, and international, with the possibility that some standards are adopted at multiple levels.

National standards are developed by a national standards body or one of its member organizations, and can be adopted at local, regional, or national levels. International standards are developed following the ISO/IEC directives standards adopted at the national level. Local standards are standards adopted at city or county levels. Regional standards are standards adopted at the state level.

STANDARDS, CODES, AND REGULATIONS

Standards represent the minimum requirements or specifications that materials, products, processes, or services should meet to ensure quality and safety. In the United States, the voluntary consensus process for developing standards means that compliance with standards is not mandatory.

A *code* may be an industry, government, or voluntary consensus-based standard or a group of standards adopted into law by a local, regional, or national authority or included in a business contract. A code represents a set of rules that serve as generally accepted guidelines recommended for the industry to follow and refers to "practices or procedures for the design, manufacture, installation, maintenance or utilization of equipment, structures or product" [2]. A code "can include references to standards, which means the standards are incorporated by reference and therefore are part of the code and legally enforceable" [5]. Well-known examples of codes include the National Electrical Code (NFPA 70) or the National Standard Plumbing Code (NSPC).

The model code is a special type of code developed with the intent of creating an industry-wide standard that can be adopted and customized by local jurisdictions. A model code becomes enforceable only after it is adopted by a jurisdiction. An example of a model code is the International Building Code (IBC) used in construction.

A *regulation* is a legally binding document adopted by a government body. A regulation may incorporate standards, codes, or be developed on its own by a government body and does not necessarily require consensus.

STANDARDS CLASSIFICATION

The following classification categories and definitions represent some of the most commonly used types of standards. Classification categories are not mutually exclusive as standards can simultaneously have characteristics of various categories.

MEASUREMENT STANDARDS

Fundamental to any human activity are the measurement standards that represent the fundamental reference to a system of measurement units for weight, length, time, temperature, and volume. These standards are the foundation for all types of human activity. Well-known examples of these standards systems are the International Metric System, which is widely adopted worldwide, and the Imperial System, which is used mostly by countries formerly under British colonial rule.

Due to its critical role in society, the science of measurement or metrology establishes a common understanding of measurement units and ensures their correct use in practice. Metrology has three levels of standards: *primary, secondary*, and *working standards*, where the *"primary standards* do not reference any other standards, the *secondary standards* are calibrated with reference to a primary standard, and the *working standards* are calibrated with respect to secondary standards" [6].

Each country has its own organization responsible for preserving the quality of these standards, and they work together at the international level to ensure uniformity of the measurement units worldwide. In the United States, the body responsible for the development, maintenance, and dissemination of measurement standards is the National Institute of Standards and Technology (NIST).

DOCUMENTARY STANDARDS

NIST defines documentary standards as "written agreements containing technical specifications or other precise criteria that may contain rules, guidelines, or definitions of characteristics" [7].

ISO identified eight types of documentary standards based on their purpose [2]; however, not all SDOs publish all eight types of standards depending on the needs of the field they serve.

STANDARDS BY ECONOMIC SECTOR

Industry Standards

Industry standards represent the minimal accepted requirements followed by the members of an industry with the purpose to ensure basic quality and safety expectations. These standards are developed by organizations representing the members of the industry and vary between different industries.

Industry standards may include any of the documentary standards, have different levels of geographical recognition, and various legal status. In the United States, most industry standards are voluntary while other countries have their own systems.

TABLE 2.1. *Types of Standards as Defined by ISO [2].*

Basic standards	Standard that contains general provisions for one particular field.
Terminology standards	Standards that define the terminology to be used in a specific field, sometimes accompanied by explanatory notes, illustrations, examples, etc.
Testing standards	Standards that prescribe the test methods and other related provisions such as sampling, use of statistical methods, sequence of tests.
Product/ component standards	Standards that specify the requirements for a product or component in order to ensure safety, interoperability, and consistency.
	Product standards include design or performance standards, where the design standards specify the design or technical characteristics of a product, and the performance standards specify the expected level of performance for a product. The connection between these standards is that the design standards are used to meet the performance goals established by the performance standards [4].
	Design standards include standards that establish the physical, chemical, electrical, and mechanical characteristics of materials and components, define product dimensions, or establish functional parameters. These standards are prescriptive in nature and may result in limited flexibility for a company.
	Performance standards include product specifications, methods and testing standards, and quality validation standards. These standards are incorporated by choice and allow for flexibility on how to reach the goals. Where feasible, performance standards are preferred to design standards because they allow for creativity and innovation.
Process standards	Standards prescribe the requirements to be met by a process in order to function effectively and safely.
Service standards	Standards that establish requirements to be met by a service in order to achieve its intended purpose effectively.
Interface standards	Standards that specify requirements for the compatibility of products or systems at their points of interconnection.
Standard on data to be provided	Standards that specify the values or data on the characteristics of a product, process, or service.

To support global economic development and facilitate international trade, there is strong interest in harmonizing industry standards across various countries.

Medical, Health, and Safety Standards

Health and safety standards are developed to help reduce workplace risks and improve the overall safety of the public. Best-known examples of these standards in the United States are the standards for patient privacy (Health Insurance Portability and Accountability Act or HIPAA), the workplace safety standards (Occupational Safety and Health Administration or OSHA), and the many regulations for food, drug, and medical devices from the Food and Drug Administration (FDA). All these standards have legal power and compliance is mandatory.

The increase in the use of information technology for health care resulted in the development of an increased number of standards for clinical data exchange or for the handling, storing, and transmitting of digital information. The governmental institution responsible for the promotion of nationwide standards-based health information technology is the Office of the National Coordinator for Health Information Technology (ONC), which is part of the U.S. Department of Health and Human Services (HHS) [8].

By complying with regulations and adopting applicable voluntary standards, health care organizations can reduce costs, accelerate the integration of new technologies, secure interoperability with existing systems, and improve quality services. For additional information on standards used in the health sciences, please refer to Chapter 13.

Food Safety Standards

Foodborne illnesses constitute a major burden on public health, and the consensus is that these illnesses could be prevented by taking measures to ensure the safety and quality of the food and food industry. Food safety standards represent the requirements that foods or food processors must comply with to safeguard human health and are implemented by authorities and enforced by law [9].

In the United States, food safety relates to the processing, packaging, and storage of food to prevent foodborne illness and is regulated by three federal governmental agencies, including the Food Safety and Inspection Service (FSIS) which is part of the U.S. Department of Agriculture (USDA), the U.S. Food and Drug Administration (FDA), and the Centers for Disease Control and Prevention

(CDC), with both the FDA and CDC functioning within the U.S. Department of Health and Human Services [10]. To ensure food quality, the FDA requires that food ingredients meet the specifications of the Food Chemicals Codex. The Food Chemicals Codex is a compendium of internationally recognized standards for the identity, purity, and quality of food ingredients [11].

The regulation of food safety started after 1906 with laws being written in response to severe outbreaks. Starting with the Food Safety Modernization Act (FSMA) of 2011, however, food safety laws had a major shift by focusing on the prevention of foodborne illnesses rather than responding as issues occurred [12].

Information and Communication Technologies Standards

With the information technology sector being critical to everyday activities all over the world, it is important that Information and Communications Technologies (ICT) organizations comply with standards developed to enable interoperability and compatibility between various ICT systems and to ensure the efficiency and security of ICT systems. ICT standards include software and hardware standards, privacy and cybersecurity standards, and Internet standards, as well as standards for the management systems that control and mitigate the risk associated with data and information.

ICT standards are developed based on the OpenStand principles that encourage "the development of market-driven standards that are global and open—enabling standards without borders and driving innovation for the benefit of humanity" [13]. Open standards are publicly available and have various use rights associated with them based on the accompanying license terms, and may or may not be available free of royalty fees. These standards are developed by internationally recognized standards bodies such as the IEEE Standards Association, Internet Engineering Task Force (IETF), International Organization for Standardization (ISO), International Electrotechnical Commission (IEC), International Telecommunication Union–Telecommunications (ITU-T), or international consortia such as the World Wide Web Consortium (W3C) and Organization for the Advancement of Structured Information Standards (OASIS).

Software standards are concerned with all processes associated with the software life cycle. The main goal of software standards is to enable interoperability between software and across platforms. It is important that software standards are implemented correctly to avoid the need for customized interfaces.

Hardware standards specify the hardware requirements necessary to ensure that the components are interchangeable and compatible with the software.

Privacy and security standards have been developed to support continuous innovation while ensuring the interoperability of data exchange and facilitating widespread adoption.

Energy Management Standards

Energy management standards help cut energy consumption and improve the security of energy systems. In the United States, these standards are the responsibility of the Department of Energy (DOE), which adopted the ISO 50001 series. The ISO 50001 is "a voluntary global standard for energy management systems in industrial, commercial and institutional facilities" [14]. DOE collaborates closely with other governmental agencies and organizations to develop policies and regulations related to energy technologies [15].

Environmental Management Standards

Environmental management standards are developed to help companies and organizations reduce their environmental impacts, reduce waste, and be more sustainable. In the United States, the Environmental Protection Agency (EPA) was established to ensure environmental protection by consolidating various federal research, monitoring, standards-setting, and enforcement activities in one governmental agency [16]. The EPA creates and enforces regulations that cover a range of environmental and public health protection issues, from setting standards for clean water to specifying cleanup levels for toxic waste sites to controlling air pollution from industry and other sources.

At the international level, ISO developed a family of standards for an environmental management system to help organizations improve their environmental performance [17].

Quality Management Systems Standards

Quality management systems standards are voluntary codes, guidelines, or processes used by organizations to formalize, systematize, and legitimize a very diverse set of managerial activities or tasks.

The ISO 9000 standards family is well known worldwide and can be used by any type of organization. The ISO 9001 standard—the best known in this family—sets out the criteria for a quality management system and ensures high-quality products and services in a consistent manner [18].

Military Specifications and Standards

Military standards and specifications are applicable to the military operations, services, and the vendors that interact with the military organizations. In the United States, these standards are developed or adopted by the Department of Defense (DoD). The aim of these standards and specifications is to streamline procurement and maintain high quality and security levels.

FUNCTIONAL STANDARDS

Allowing innovation implementations without affecting the expected functionality of a system is critical, especially when the system is dependent on information technologies that require regular updating and improvement. This flexibility can be achieved by setting functional standards that separate system functionality from the technology or process(es) that are part of the system, and focus only on what the system is expected to do and the capabilities it needs to have.

Functional standards are different from *performance standards* as the first defines the minimum operational requirements for a system while the second defines the metrics by which the system will be evaluated.

SUMMARY

Standards and regulations play a critical role in the global economy and society by providing the guidelines and best practices that ensure safety, interoperability, and quality for products, processes, and services. Standards can be classified by different criteria, but the classification categories are not mutually exclusive. Compliance with standards is voluntary except when the standards were adopted into law by a local, regional, or national authority or were included in a business contract.

REFERENCES

1   ISO COPOLCO, "Standards in our world," *Consumers and standards: Partnership for a better world*. https://www.iso.org/sites/ConsumersStandards/1_standards.html (accessed Apr. 29, 2021).

2   "ISO/IEC guide 2:2004 (2016) (multilingual) standardization and related activities— general vocabulary," 2016. https://isotc.iso.org/livelink/livelink/fetch/2000/2122 /4230450/8389141/ISO_IEC_Guide_2_2004_%28Multilingual%29_-_Standard ization_and_related_activities_--_General_vocabulary.pdf?nodeid=8387841& vernum=-2 (accessed Apr. 29, 2021).

3   "OMB circular A-119, revised: Federal participation in the development and use of voluntary consensus standards and in conformity assessment activities," 2016. https://www.nist.gov/system/files/revised_circular_a-119_as_of_01-22-2016.pdf (accessed Apr. 29, 2021).

4   "Circular NO. A-119 revised," 2017. https://www.whitehouse.gov/wp-content /uploads/2017/11/Circular-119-1.pdf (accessed Apr. 29, 2021).

5   M. Heinsdorf, "Code or standard?," *Consulting-Specifying Engineer*, July 1, 2015. https://www.csemag.com/articles/code-or-standard

6   G. M. S. de Silva, *Basic metrology for ISO 9000 certification*. Routledge, 2012.

7   M. A. Breitenberg, "The ABC's of standards activities," National Institute of Standards and Technology, Gaithersburg, MD, NIST IR 7614, 2009. https://doi.org /10.6028/NIST.IR.7614.

8   Office of the National Coordinator for Health Information Technology, "Health IT standards." https://www.healthit.gov/topic/standards-technology/health-it-stand ards (accessed Jul. 17, 2021).

9   ISO, "ISO 22000—food safety management." https://www.iso.org/iso-22000-food -safety-management.html (accessed Jun. 30, 2021).

10 "About FoodSafety.gov," May 23, 2019. https://www.foodsafety.gov/about (accessed Jun. 30, 2021).

11 "Food Chemicals Codex (FCC)." https://www.foodchemicalscodex.org (accessed May 29, 2022).

12 U.S. Food and Drug Administration, "Food Safety Modernization Act (FSMA)," Jun. 11, 2021. https://www.fda.gov/food/guidance-regulation-food-and-dietary -supplements/food-safety-modernization-act-fsma (accessed Jun. 30, 2021).

13 OpenStand, "The modern standards paradigm—five key principles." https:// open-stand.org/about-us/principles (accessed Jun. 30, 2021).

14 "DOE Updates ISO 50001 Programs to New Standard." https://www.energy.gov /eere/amo/articles/doe-updates-iso-50001-programs-new-standard (accessed Jul. 15, 2021).

15 U.S. Department of Energy, "Supplemental information: Agency information," in *Quadrennial Technology Review 2015*. https://www.energy.gov/sites/default/files /2016/02/f29/Ch.1-SI-Agency-Information.pdf (accessed Jul. 15, 2021).

16 U.S. Environmental Protection Agency (EPA), "About EPA," Jan. 18, 2013. https:// www.epa.gov/aboutepa (accessed Jul. 15, 2021).

17 ISO, "ISO 14000 family—environmental management." https://www.iso.org/iso -14001-environmental-management.html (accessed Jul. 15, 2021).

18 ISO, "ISO 9000 family—quality management." https://www.iso.org/iso-9001 -quality-management.html (accessed Jul. 15, 2021).

3

Development through the Standardization Process

Erin M. Rowley, University at Buffalo (SUNY)

INTRODUCTION

There are many reasons to create a standard, as discussed in previous chapters; however, the next logical questions that may arise are: How does one go about creating a standard? Can anyone suggest or create a standard? Who needs to be involved? Is a standard created once and then "written in stone," as the saying goes, or does it need to be updated? If it is updated, who is responsible for that piece? As with questions related to standards, the answers to these questions depend on a variety of factors.

Standards are created by standards developing organizations (SDOs). SDOs can be, but are not limited to, national organizations, international organizations, or trade organizations. Even companies can create their own standards, so they too could be considered an SDO. Each SDO determines the process for creating, reviewing, and revising its standards. This chapter will explore the standardization development process for several different U.S.-based organizations, as well as how select countries and international organizations approach the process. In addition, the chapter will explore how governmental standards are developed and how open standards are developed and maintained. Lastly, this chapter will examine how standards and standards updates are disseminated to stakeholders and the public at large.

HOW "VOLUNTARY CONSENSUS" STANDARDS ARE DEVELOPED

Voluntary consensus standards are "technical specifications for products or processes that are developed by standards-setting bodies" [1]. Standards-setting bodies include SDOs like ASTM International, the American Society of Mechanical Engineers (ASME), the Society of Automotive Engineering (SAE), and the Technical Association of the Pulp and Paper Industry (TAPPI). It is most common for standards-setting bodies to create standards by consensus of industry and technical experts. As the National Resource Council stated:

> These groups write standards through a formal process of discussion, drafting, and review. Group members attempt to form consensus on the best technical specifications to meet customer, industry, and public needs. The resulting standards are published for voluntary use throughout industry [2].

The actual process and timeline for developing standards can vary depending on the SDO; however, many of these organizations share the process guidelines they adhere to during the standards development process on their websites. Figure 3.1 shows an example of the typical process for developing a voluntary consensus standard. This is only meant to provide a general example as to how a standard might be created. Some SDOs may require multiple comment periods from the public. Others may require a vote from members before a standard is finalized.

Figure 3.1 illustrates when a new standard is created, but the process may look similar for existing standards, where an SDO determines if any changes need to be made to the current document, or if the standard can be "reaffirmed," or approved for another set term without any changes [3].

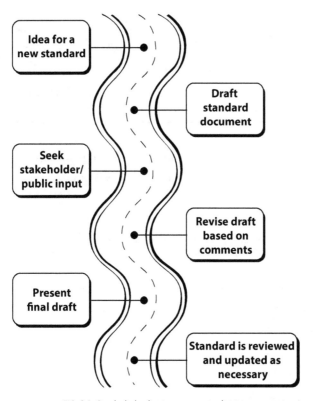

FIG. 3.1. Standards development process visualization.

The technical experts that make up these SDOs can come from many different areas including industry and academia. Experts may work for those companies that are key stakeholders for the standard in question. For example, toy design engineers from a major toy company may sit on the ASTM subcommittee on toy safety, which is responsible for ASTM F963 Standard Consumer Safety Specification for Toy Safety [4]. In addition, other experts in the field, such as test engineers and product specialists from third-party testing laboratories, may sit on committees within SDOs. Third-party laboratories conduct testing to certain standards. In some cases, third-party testing is required under the law to prove compliance with a certain standard.

As standards are technical documents, most SDOs aim to have their standards follow a particular style template when new standards are created. SDOs create their own templates and style guidelines, with many of these guidelines freely accessible online. See Table 3.1 for a small sampling of SDO style guidelines. Also, refer to the further reading section for additional information.

TABLE 3.1. *List of Standards Developing Organization Style Guidelines*

ASTM	https://www.astm.org/FormStyle_for_ASTM_STDS.html
ISO house style	https://www.iso.org/ISO-house-style.html
TAPPI	https://www.tappi.org/content/pdf/standards/tm_guidelines _complete.pdf

An Internet search for the SDO and the phrase "style guideline" can typically bring back results on templates and rules for creating a standard to be associated with that SDO.

HOW OTHER STANDARDS ARE DEVELOPED

Beyond voluntary consensus standards, two other general categories of standards exist, namely *de facto standards* and *mandatory standards* [2] as was touched upon in Chapter 2, "An Exploration of Types of Standards."

De facto Standards

De facto standards are private-sector standards, similar to voluntary consensus standards developed through trade and industry groups, however, they are not formed via consensus. More specifically, de facto standards are those that have "become accepted in practice but [have] not undergone any formal process to obtain consensus and may not even have publicly available documentation" [5]. The term "de facto" is Latin for "in fact." In the case of standards, de facto refers to that the standards are followed but not necessarily mandatory under the law.

Mandatory Standards

Conversely, *mandatory standards* are, as the name implies, standards that are mandatory under a law or regulation. In the United States, as an example, standards are traditionally voluntary unless they are made mandatory under federal, state, county, or city law or regulation. Of course, this is complicated by the practice of using the term "standard" for a variety of different documents. For example, the Federal Motor Vehicle Safety Standards (FMVSS) issued under the

National Highway Traffic Safety Administration (NHTSA) use the term "standards" in the title of the document, however, these are mandatory as the "standards" themselves are actually regulations under Title 49 of the U.S. Code of Federal Regulations [6].

STANDARDS DEVELOPMENT IN THE UNITED STATES COMPARED TO OTHER COUNTRIES

Just as standards development can vary from one SDO to the next, standards development is different in the United States versus other countries or regions, such as the European Union, China, South Korea, and Australia. There are certainly many intricacies when it comes to the standards development process, therefore, this section is intended to provide a brief overview of these selected countries and geographic regions.

Standards Development in the European Union

European standards are developed by the European Standardization Organizations: the European Committee for Standardization (CEN), the European Committee for Electrotechnical Standardization (CENELEC), and the European Telecommunications Standards Institute (ETSI). The 34 member states of the European Union work together to develop standards (among other deliverables); more than 50,000 technical experts from industry, academia, trade associations, public administrations, and societal organizations are involved in the network to produce and maintain these technical documents [7].

The development of a European Standard (EN) is done via a multistage process, including a proposal, drafting, public comment, and formal vote. CEN states the process of creating an EN standard "is the result of a transparent, open and consensual development process with national commitment" [8].

Please note, individual member states of the European Union have their own national standards bodies (NSBs) that adopt and publish national standards. NSBs may have slightly different policies and procedures for creating and revising national-level standards. NSBs also "transpose all European standards as identical national standards and withdraw any conflicting national standards"

[9]. These standards often become mandatory when cited in a European Union directive (e.g., electromagnetic compatibility [EMC] and personal protective equipment [PPE] standards).

Standards in China

The Standardization Administration of the People's Republic of China (SAC) is the primary national technical standards agency in China. Established in 2001, the SAC "is authorized by the State Council to undertake unified management, supervision and overall coordination of standardization work in China" [10]. The SAC states the full responsibilities of their organization are to:

- Deliver national standards plans, approve and publish national standards, deliberate and release important documents such as standardization policies, administrative rules, programs and announcements;
- Notify mandatory national standards to the public;
- Coordinate, guide and supervise standards work concerning industry, local areas, organizations and enterprises;
- Represent China to join ISO, IEC and other international or regional standardization organizations;
- Sign and execute international standards cooperation agreements; and
- Undertake the daily work of the standardization coordination mechanism under the State Council [11].

In October 2021, the Chinese government released information regarding a new technical standards strategy, the National Standardization Development (NSD) Outline, also known as China's 2035 Standards Plan [12, 13]. China's plan for technical standards differs from other countries. As the Carnegie Endowment for International Peace noted, in China, "standards are often seen as a lever for upgrading the country's industrial base." The 2035 Standards Plan "focuses heavily on creating standards for emerging industries such as intelligent manufacturing," [14] describing them as a tool to "promote industry optimization and upgrading" [12]. This difference is important to note, especially as more businesses and industries expand internationally.

Standards Development in South Korea

The standards process in South Korea differs slightly from those in Europe and North America, as Korea's standards development is still largely led by the legislative process, specifically under the Industrial Standardization Act of 1961 [15]. Making things more complicated are the relatively few resources that exist in English regarding standards in Korea.

Simply put, the Korean Agency for Technology and Standards (KATS) is the governmental agency in Korea under the Ministry of Trade, Industry, and Energy (MOTIE) and develops Korean Industrial Standards (KS) [16]. There are a number of laws created under KATS in the KS system, including the Framework Act on National Standards (1999), the Measures Act (1961), and the Electrical Appliances Safety Control Act (1974) [15].

The Korean Standards Association (KSA) is then responsible for the sale of Korean standards. The KSA also is responsible for standards education in South Korea, which includes from the elementary school level to the university/college level. Education content also is created for industries and general consumers [17].

Standards Development in Australia

In Australia, standards are mostly developed through Standards Australia, or jointly through Australia/New Zealand Standards. Standards Australia details a six-stage process for developing an Australian standard, which includes a public comment period and a ballot vote by the Standards Development and Accreditation Committee (SDAC) [18]. Any proposal to create a new standard or amend an existing standard comes directly from the Australian community [19].

Similar to the United States, these standards are considered voluntary until they are made mandatory under Australian laws and regulations. Interestingly, it is common for lawmakers in Australia to make only certain portions of a standard mandatory under a regulation, such as with the safety standard for bicycles, AS/NZS 1927:1998 [20]. In this case, only portions of the standard, called out by Australian Consumer Protection Notice No. 6 of 2004, are considered mandatory [21].

GOVERNMENT STANDARDS

National standards bodies (NSBs) can also create standards; however, these can be different from government standards. In the United States, the American National Standards Institute (ANSI) is widely regarded as one of the country's primary NSBs; however, U.S. governmental agencies, which are not affiliated with ANSI, can create their own standards as well. The U.S. General Services Administration (GSA) provides the *Index of Federal Specifications, Standards, and Commercial Item Descriptions* (FMR 102-27) [22], which includes a listing of hundreds of federal standards [23]. These documents are separate from U.S. Military Standards, which are "documents developed and used for products, materials, and processes that have multiple applications to promote commonality and interoperability among the Military Departments and Defense Agencies" [24].

Government standards are developed differently than voluntary consensus standards. Military standards are maintained by the Defense Supply Center Columbus (DSCC), part of the Defense Logistics Agency. The Department of Defense also works with private contractors to produce material [25].

Other examples of U.S. government standards include those from other governmental agencies such as the Environmental Protection Agency (EPA), the Federal Aviation Administration (FAA), the National Aeronautics and Space Administration (NASA), the Nuclear Regulatory Commission (NRC), and the Occupational Safety and Health Administration (OSHA).

OPEN STANDARDS

Open standards are standards that are made available to the public [26]. Typically, when the term "open standards" is used, many familiar with standards think of the Internet or other information technology standards that exist. The Internet Society states:

> the Internet is fundamentally based on the existence of open, non-proprietary standards. These standards are key to allowing devices, services, and applications to work together across a wide and dispersed network of networks [27].

The Internet Society sponsors the RFC (Request for Comments) Editor website that contains a listing of technical and organizational documents about the

Internet. A listing of official Internet protocol standards can give some context to how many open standards exist for the Internet [28].

There are many reasons for standards to be open [29]. It is important to note that "open" does not always equate to "free," but freely available standards exist. For example, the National Operating Committee on Standards for Athletic Equipment (NOCSAE) provides their standards without charge via their website. However, NOCSAE standards would not be considered "open" as they are copyrighted documents [30]. This is a departure from many other trade- and industry-based SDOs. In the United States, other standards are available without charge if they are specifically listed or named in federal laws or regulations, known as incorporated by reference. This topic is covered in more detail in Chapter 5, "Discovering and Accessing Standards," and Chapter 6, "Standards Collection Development."

CONFORMITY ASSESSMENT AND CERTIFICATION

Standards are used to create consistency among products and processes, but it is conformity assessment that ensures compliance with said standards. Per ISO, conformity assessment "involves a set of processes that show your product, service or system meets the requirements of a standard" [31]. This benefits all key stakeholders as it provides consumers with confidence in the product, it can give the producing company a competitive edge by showing compliance with a standard or standards, and it helps regulatory bodies ensure all key requirements related to health and safety were met.

HOW STANDARDS ARE DISSEMINATED

Standards are disseminated in a variety of ways, especially with the help of the Internet with more and more information being distributed digitally. Individual SDOs typically create their own means of disseminating information related to their standards, including standards that have been updated or newly created standards. For example, ASTM International states that they utilize a variety of means to communicate to their stakeholders about standards, including using their Digital Library database, training courses, enterprise solutions, proficiency testing, certification and declaration, and the *Standardization News* magazine [32].

As previously mentioned for South Korea, some SDOs consider the training of students, industry, and the general public to be a major portion of their role. KSA provides standards education through regular on-site programs, as well as e-learning options. Further, they publish reports and books on standards development and attend and arrange conferences, seminars, and forums [33].

Social media also is used heavily by most SDOs. In the United States, SAE, ASME, ASCE, ASTM, and many others have extremely active newsfeeds on platforms including Twitter, Facebook, and LinkedIn. In turn, these social media feeds can be picked up by both national and local media channels, typically depending on the topic of the standard. Media outlets also help create broader awareness for standards when they report on product recalls, such as those related to furniture tip over [34]. The standard ASTM F2057, "Standard Safety Specification for Clothing Storage Units" dictates requirements to prevent the tip over of these large furniture items, which have caused injury and death to children attempting to climb nonanchored furniture.

SUMMARY

Like many things related to standards, there is no "one answer" for the standardization process. It can vary greatly depending on the type of standard being created and the organization or agency the standard falls under; however, as standards are technical documents, some generalities can be applied. Standards may not always be continually updated, but they typically are intended to be living documents that are reviewed and adjusted as necessary. They are developed to create consistency or a guideline for others to follow.

REFERENCES

1   *Voluntary Consensus Standards: Agencies' Compliance With the National Technology Transfer and Advancement Act. Statement of Jill Wells, Director, Energy Resources, and Science Issues, Resources, Community, and Economic Development Division.* Testimony before the Subcommittee on Technology, Committee on Science, House of Representatives, 2000. https://www.gao.gov/assets/t-rced-00-122 .pdf (accessed November 27, 2020).

2   National Research Council, International Standards, C. A., and Committee, U. S. T. P. P. Standards, conformity assessment, and trade into the 21st century, Interna-

tional Standards, Conformity Assessment, and U.S. Trade Policy Project Committee, Board on Science, Technology, and Economic Policy. National Academy Press, 1995.

3   SAE International, "Standards Status Definitions." https://www.sae.org/standards /development/definitions (accessed August 2, 2021).

4   ASTM International, "ASTM F963 Toy Safety Update." https://www.astm.org/toys .html (accessed November 27, 2020).

5   R. Campbell, E. Pentz, and I. Borthwick, *Academic and professional publishing*. Oxford: Chandos Publishing, 2012.

6   Federal Motor Vehicle Safety Standards, 49 C.F.R. § 571, National Highway Traffic Safety Administration (NHTSA), Department of Transportation (DoT), 2021. https:// www.ecfr.gov/current/title-49/subtitle-B/chapter-V/part-571 (accessed November 20, 2020).

7   European Committee for Standardization (CEN), "CEN and CENELEC." https:// www.cencenelec.eu/european-standardization/cen-and-cenelec (accessed November 27, 2020).

8   European Committee for Standardization (CEN), "European Standard (EN)." https:// boss.cen.eu/developingdeliverables/pages/en/pages (accessed October 26, 2021).

9   Your Europe, "Standards in Europe." https://europa.eu/youreurope/business /product-requirements/standards/standards-in-europe/index_en.htm (accessed November 27, 2020).

10 Standards Administration of the P.R.C., "Who we are." http://www.sac.gov.cn /sacen/aboutsac/who_we_are/201411/t20141118_169916.htm (accessed July 30, 2022).

11 Standards Administration of the P.R.C., "What we do." http://www.sac.gov.cn /sacen/aboutsac/What_we_do/201411/t20141118_169917.htm (accessed July 30, 2022).

12 Z. Ying, "The Central Committee of the Communist Party of China and the State Council issued the 'National Standardization Development Outline.'" http://www .gov.cn/zhengce/2021-10/10/content_5641727.htm (accessed November 20, 2020).

13 Y. Wu, "China Standards 2035 Strategy: Recent Developments and Implications for Foreign Companies," *China Briefing*, July 26, 2022. https://www.china-briefing.com /news/china-standards-2035-strategy-recent-developments-and-their-implications -foreign-companies

14 M. Sheehan, M. Blumenthal, and M. R. Nelson, "Three takeaways from China's new standards strategy," October 28, 2021. https://carnegieendowment.org/2021/10/28 /three-takeaways-from-china-s-new-standards-strategy-pub-85678

15 D. G. Choi, "A primer on Korea's standards system: Standardization, conformity assessment, and metrology," National Institute of Standards and Technology (NIST), U.S. Department of Commerce, January 2013 2013. https://nvlpubs.nist.gov

/nistpubs/ir/2013/NIST.IR.7905.pdf

16 Korean Agency for Technology and Standards (KATS), "Korean industrial standards." https://kats.go.kr/content.do?cmsid=27 (accessed November 27, 2020).

17 Korean Standards Association (KSA), "Standardization and standards." https://eng.ksa.or.kr/ksa_english/5176/subview.do (accessed November 27, 2020).

18 Standards Australia, "Our process." https://www.standards.org.au/standards-development/developing-standards/process (accessed November 27, 2020).

19 Standards Australia, "Submitting a proposal." https://www.standards.org.au/standards-development/developing-standards/proposal (accessed November 27, 2020).

20 Product Safety Australia, "Bicycles." https://www.productsafety.gov.au/product-safety-laws/safety-standards-bans/mandatory-standards/bicycles (accessed November 27, 2020).

21 "Consumer product safety standard: Pedal bicycles: Safety requirements, Commonwealth of Australia Consumer Protection Notice No. 6 of 2004," 2004. https://www.legislation.gov.au/Details/F2005B01053 (accessed November 20, 2020).

22 U.S. General Services Administration, "Index of Federal Specifications, Standards and Commercial Item Descriptions (FMR 102-27)." https://gsafas.secure.force.com/fedspecs?CFID=6185597&CFTOKEN=4bedc9bd2f4fdb6e-0848A54 3-5056-8700-95B267BC40CA28F2 (accessed January 4, 2021).

23 U.S. General Services Administration, "List of federal standards." https://fedspecs.gsa.gov/FedSpecs_Fed_Standards_page (accessed January 4, 2021).

24 Defense Standardization Program, "Specifications and standards." https://www.dsp.dla.mil/Specs-Standards (accessed January 4, 2021).

25 M. H. Jawad, *Primer on engineering standards.* Hoboken, NJ: Wiley, 2018.

26 International Telecommunication Union, "Definition of 'Open Standards.'" https://www.itu.int/en/ITU-T/ipr/Pages/open.aspx (accessed January 12, 2021).

27 Internet Society, "Open Internet standards chapter toolkit." https://www.internetsociety.org/chapters/resources/open-internet-standards-chapter-toolkit (accessed January 12, 2021).

28 RFC Editor, "Official Internet protocol standards." https://www.rfc-editor.org/standards#IS (accessed January 12, 2021).

29 L. DeNardis, Opening standards the global politics of interoperability. Cambridge, MA: MIT Press, 2011.

30 National Operation Committee on Standards for Athletic Equipment (NOC-SAE), "Standards matrix." https://nocsae.org/standards/standards-matrix/#/all/performance/current (accessed January 13, 2021).

31 International Organization for Standarization (ISO), "Certification and conformity." https://www.iso.org/conformity-assessment.html#:~:text=Conformity%20 assessment%20involves%20a%20set,your%20company%20a%20competitive%20 edge (accessed January 13, 2021).

32 ASTM International, "ASTM fact sheet." https://www.astm.org/about/overview /fact-sheet.html (accessed January 4, 2021).

33 Korean Standards Association (KSA), "Training and education." https://eng.ksa .or.kr/ksa_english/5179/subview.do (accessed January 4, 2021).

34 K. Gibson, "Kids' advocates: Deaths and recalls show need to mandate furniture standards." https://www.cbsnews.com/news/furniture-deaths-recalls-mandate-stability (accessed April 4, 2021).

FURTHER READING

ASTM International Form and Style for ASTM Standards (April 2020): https://www. astm.org/media/pdf/bluebook_FormStyle.pdf

ETSI A Guide to Writing World Class Standards: https://www.etsi.org/images/files/ Brochures/AGuideToWritingWorldClassStandards.pdf

IEEE Write Your Standard: https://standards.ieee.org/develop/drafting-standard/ write

ISO How to Write Standards: https://www.iso.org/files/live/sites/isoorg/files/archive/ pdf/en/how-to-write-standards.pdf

PART II

Standards Access and Collection Development to Support Information Literacy

4

Determining Standards Information Literacy Needs

Daniela Solomon, Case Western Reserve University

INTRODUCTION

Information literacy as defined by the American Library Association (ALA) is "the set of integrated abilities encompassing the reflective discovery of information, the understanding of how information is produced and valued, and the use of information in creating new knowledge and participating ethically in communities of learning" [1]. The development of information literacy skills in science, engineering, technology, and professional disciplines is challenging due to the large variety of information sources available, the diversity of publishers, and the application of information within their fields. This is especially true for grey literature that is produced outside of the traditional academic publishing system. Grey literature requires a deeper understanding of this type of information along with the knowledge of the agency or organization publishing the information. Standards are considered one example of grey literature.

Technical standards embody critical information that engineering and professional practitioners are required to apply in their activities to secure the success of their enterprises. The increased globalization and worldwide trade make knowledge of standards and standardization even more crucial. In turn, developing standards information literacy is of major significance for engineering and professional students.

STANDARDS INFORMATION LITERACY

Identifying the technical standards knowledge and skill set needed by students at graduation has been of interest for more than a decade. In 2003, the Standards Education Task Force within IEEE surveyed students and faculty to identify the state of standards education in academic programs. As a result, the task force recommended that, at minimum, students should [2]:

- develop a basic understanding of the standardization process, the impact of standards, and how standards are beneficial to the global economy;
- become familiar with the key standards organization in their discipline; and
- learn how to identify relevant standards and utilize them in engineering design.

In 2009, the IEEE University Outreach Program continued the investigation into how to facilitate standards education in academia and reported [3]:

- the differences between the standards needs of undergraduates and graduate students;
- academic institutions may take various approaches to teach about standards; and
- the differences between how various countries conduct standards education.

This study confirmed the list of basic technical standards knowledge identified previously and added that students should develop an understanding of the role technology, economics, and politics play in standards development, learn to think critically about standardization, and recognize the impact standards have on innovation.

Building on these two studies, a later survey identified the benchmark practices in teaching technical standards and concluded that faculty prefer that standards education be integrated into the curriculum as coursework, assignments, case studies, lectures, industry expert visits, and so forth, and not standalone courses. The faculty also were interested in offering the students opportunities to practice standards identification, retrieval, and evaluation [4].

These recommendations align perfectly with the definition of information literacy from the Association of College and Research Libraries Information Literacy Framework:

- Understanding of how information is produced and valued
  - Understand the type of information found in standards and its value
  - Understand the standardization process
  - Understand standards benefits to the global economy and society
- Discovery of information
  - Identify standards relevant to a specific project
  - Learn about key standards developing organizations (SDOs) in their discipline
  - Learn about standards resources
  - Locate standards
- Use of information in creating new knowledge
  - Learn about the different standards sections and their purpose
  - Be cognizant of the language used in standards

A short but comprehensive introduction to technical standards that includes the definitions, types, and benefits of standards could address the understanding of how standards are produced and their value. However, developing skills in locating, evaluating, and using standards call for more training and opportunities for practice [4].

DISCOVERY OF INFORMATION

Identification of Relevant Standards

If the course requires students to use standards for a project or assignment, the first step in incorporating technical standards information literacy is to help the students identify the relevant standards. The awareness of the existence and applicability of codes and standards to a course project ensures the quality and safety of the final course product and can be passed on to stakeholders. When identifying standards, it is important to remind students that compliance with industry standards is voluntary in the United States; however, noncompliance with standards and codes may result in products being recalled, rejected, or fined by regulators, interoperability issues with products, or the inability to use completed products. Noncompliance may also result in accidents, illness, or property damage with grave consequences for companies.

During the research portion of the course project, the students identify the stakeholders' requirements and develop a solid understanding of the problem to be solved. The research stage should include searching for patents, applicable codes, and technical standards along with other types of literature. Identifying standards and codes early on in the research process can save time and money by facilitating market access and acceptance, secure production flexibility and manufacturing responsiveness, and improve quality. Standards are applicable at every stage of the engineering design process, from engineering drawings to quality management, as seen in the example in Figure 4.1.

Having students talk with stakeholders or an experienced professional in the field will provide them with a list of codes and standards applicable to a specific project. When this approach is not possible, there are several different pathways to identify the applicable standards.

One pathway is to check government agencies to identify the major applicable regulations. The Code of Federal Regulations (CFR, https://ecfr.io) lists all governmental regulations adopted in the United States. Additionally, each government agency may have its own list of regulations such as the Consumer Product Safety Commission (CPSC, (https://www.cpsc.gov/Regulations-Laws--Standards), the Federal Communications Commission (FCC), or the U.S. Food and Drug Administration (FDA). For a more comprehensive list, see the list of standards issued or adopted by federal agencies maintained by the National Institute of Technology (NIST, https://sibr.nist.gov) or the American National

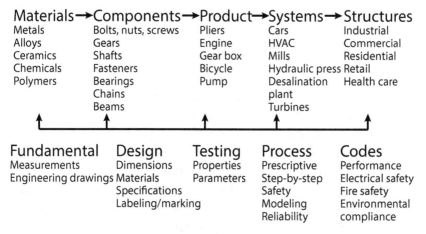

FIG. 4.1. Standards applicability to engineering design process example.

Standards Institute (ANSI, https://ibr.ansi.org). When unsure about what agency to use, a good starting point may be ANSI's Standards Packages (https://web store.ansi.org/packages/all) or Selected Standards lists (https://webstore.ansi. org/industry/selected-standards), which group relevant standards by industries. Another similar list is ANSI's Directory of Standards Organizations (https:// www.standardsportal.org/usa_en/resources/sdo.aspx). Accessing these lists may help identify the major standards organizations' publishing standards for a specific industry. Then, applicable standards can be identified by checking the list of standards published by that specific SDO.

For standards that are international in scope and participation, a good starting point is searching one of the well-known international SDOs: the International Organization for Standardization (ISO, https://www.iso.org), the International Electrotechnical Commission (IEC, https://www.iec.ch/homepage), or the International Telecommunication Union (ITU, https://www.itu.int/en/Pages /default.aspx), as applicable to the product in question.

Military standards can be found using ASSIST Quick Search (https://quick search.dla.mil/qsSearch.aspx), a searchable database of full-text defense and federal specifications and standards published by the Department of Defense.

Another pathway is to do a topic or keyword search on the Internet or use one of the third-party standards stores such as Techstreet, IHS Markit, SAI Global Standards Infobase, or ANSI webstore. These platforms are free to access and search, and will provide a short description of the content to help readers determine the usefulness of a particular standard. However, access to a standard's full text is possible for a fee. When searching using keywords, it is always good practice to consider the synonyms or phrases relevant to the product and build effective strategies to refine your search.

Yet another pathway to finding technical standards is to check for similar products and identify standards used in the design process. This pathway may not result in a comprehensive list of applicable standards, but it will give an idea of where to start. The selection of applicable standards for a product is complicated by overlaps between various SDOs that may develop standards independently [5].

When unsure where to start when identifying technical standards, students should reach out to their engineering librarians/information specialists at their institutional library or the course faculty/instructors. A tool that can be helpful for students is an information guide (LibGuide) that includes links to institution-specific technical standards collections and can be a resource for students throughout their projects.

Access to Standards

The next step when incorporating information literacy into a course is teaching the different ways technical standards are accessed. The first sources to check for standards access are the library catalogs and resources, either in print or electronic format. Libraries may also offer on-demand purchasing services for standards with Techstreet or IHS, allowing individual access to selected standards based on price and availability.

Some standards organizations give free read-only or limited access to their standards. The NIST maintains the list of these organizations and is made available on the www.standards.gov portal.

Standards incorporated by reference in the U.S. Code of Federal Regulations are available as free read-only and listed in specialized databases: ANSI Incorporated by Reference (IBR) Portal (https://ibr.ansi.org) and NIST Standards Incorporated by Reference (SIBR) Database (https://sibr.nist.gov).

If the full text is still not found, the last resort is the standards developing organization website or the authorized standards stores since most standards may be purchased in electronic form or hard copy from these sites. For additional information on accessing standards, please refer to Chapter 5.

USE OF INFORMATION IN CREATING NEW KNOWLEDGE

Once technical standards have been identified and located, the next step is to help the student or patron to understand the information included in this information type. The application of the knowledge included in the standard should be interpreted by the individual and their specific project needs. Standards have a unique structure and language characteristics intended to help understand the requirements and specifications included in the text and aid with its implementation. Therefore, preexisting knowledge of the structural elements and language characteristics facilitates understanding.

Standards follow a clear structure that makes it easy to distinguish between the various informative and normative sections; however, it is important to note that the exact structure can vary among standards developing organizations. The informative elements provide the context and scope along with additional components that help determine the applicability of a standard. Normative elements are written to eliminate ambiguity, and include the requirements to be implemented and provide the details of what exactly needs to be done when implementing the standard. Documents include verbal forms that allow for easy identification of the requirements mandated and distinguish these from other types of recommendations, permissions, possibilities, and capabilities. To avoid misinterpretations of technical standards, only the widely accepted verbal forms are used; however, both standards structure and language are dynamic, and changes are made over time [6].

STRUCTURE OF A STANDARD

Informative Elements

Designation and title—indicative of the issuing SDO, year of adoption or revision, and the applicability of a standard. A standard is recognized by its designation, which includes the acronym of the issuing body, a number and date of publication, and a title. For example: ASTM F1568 - 08 Standard Specification for Food Processors, Electric. In this example, the issuing body is the ASTM. The "08" at the end of the number refers to the date of adoption or revision, and it is followed by the title of the standard.

Foreword—describes the content of the standard and gives information on the standard development. It also details the developing process, stating the organization responsible for publishing the document, the committee that developed the document, the procedures and rules under which the document was developed, the voting process, legal disclaimers, and relationships between the present document and other documents.

Introduction—provides specific information or commentary about the document's technical content and the reasons for its preparation. This section is optional.

Scope—provides a succinct and factual statement of the document's purpose. This may include what the document does (specifies, establishes, gives guidelines, defines, provides), the subject of the document, and the aspects covered. This section is important to understand before continuing to read the standard as it helps determine the applicability of a standard or particular parts of it to the product at hand.

Normative references—list of referenced standards, where some or all their content constitutes requirements for the document. Information on how these references apply is found in the place where they are cited in the document and not in the normative references clause.

Terms and definitions—provide definitions necessary for the understanding of certain terms used in the document.

Normative Elements

Clauses and subclauses—serve as the basic components of the content/what needs to be implemented, all the details.

VERBAL EXPRESSIONS

Language is critical to the unambiguous understanding of the provisions in a standard. Terminology definitions included in the informative section ensure a common understanding of the main concepts used in the document. Verbal hints used throughout the entire content further help to identify and distinguish the provisions of a standard.

The provisions in a standard are an "expression in the content of a normative document that takes the form of a statement, an instruction, a recommendation, or a requirement" [7]:

Requirements do not allow deviation from the claims in the standard. To facilitate easy identification, requirements use verbal expressions such as "shall" or "shall not."

Recommendations include the expressions "should," "should not," or a similar phrase to suggest that the claim allows for some degree of deviation.

Permissions includes the expressions of "may" to convey consent or liberty to do something.

Possibility and capability include the expressions "can" or "cannot." Possibility refers to the expected outcomes and/or qualities. Capability refers to the ability to do or achieve a specified thing.

SUMMARY

Studies identified technical standards education topics for undergraduate and graduate students. These topics align with the information literacy concepts; however, as the process of identification, access, and effective use of relevant standards is more complex and more difficult than for other types of information, students benefit from technical standards education integration into the academic curriculum.

REFERENCES

1 Association of College and Research Libraries, "Framework for information literacy for higher education," 2015.

2 A. S. Khan, A. Karim, and J. A. McClain, "The state of the use of standards in engineering and technology education," presented at the *ASEE Annual Conference and Exposition, Conference Proceedings*, 2013.

3 B. Harding, "Lessons from professors: What the IEEE learned from global university outreach," presented at the *ICES 2011 Workshop*, Hangzhou, China, Jun. 27, 2011. https://www.standardsuniversity.org/course/lessons-from-professors-what-the-ieee-learned-from-global-university-outreach (accessed Dec. 15, 2021).

4 S. I. Rooney and J. S. Stephens-Epps, "Incorporating engineering standards throughout the biomedical engineering curriculum," presented at *the ASEE Annual Conference and Exposition, Conference Proceedings*, 2019.

5 A. S. Khan and A. Karim, "Importance of standards in engineering and technology education," *International Journal of Educational and Pedagogical Sciences, 10(3),* 2016, pp. 1050–1054.

6 N. Abdelkafi, R. Bolla, C. J. Lanting, A. Rodriguez-Ascaso, M. Thuns, and M. Wetterwald, *Understanding ICT standardization: principles and practices,* ETSI, 2018.

7 ISO/IEC, "ISO/IEC directives, part 2, principles and rules for the structure and drafting of ISO and IEC documents," 2021.

5

Discovering and Accessing Standards

Margaret Phillips, Purdue University

INTRODUCTION

Standards are specialized resources that present many search and access challenges [1]. This chapter summarizes the common approaches to searching and accessing standards currently used by academic libraries, highlights many of the standards-related challenges faced by librarians and users, and shares a listing of standards that are freely available in full text, either openly or for educational purposes.

Search Engines

There are several search engines that can be used for standards discovery and access, including some freely available options. The search engines can be categorized in four ways. This section provides more information on each of these categories:

- Standards aggregator databases;
- Standards developing organization databases;
- Library catalogs/discovery layers; and
- Other library databases.

Standards Aggregator Databases

Standards aggregators are organizations that sell standards from many different standards developing organizations (SDOs). In general, standards aggregators do not create standards themselves, but provide databases for customers to locate and purchase standards documents from a wide variety of SDOs. Many of the aggregators offer both free and paid search options. Typically, the free search options allow users to search for and browse standards and provide descriptive information, such as the document title, abstract/scope, cost, and number of pages; however, the free search options do not permit access to the full text without payment in most cases. Most of the aggregators give options for purchasing single-use PDFs and/or hard copies of individual standards. Some aggregators provide previews of a limited number of pages of selected full-text standards with their free search option. The aggregator's paid search options give access to subscription platforms with premium features, such as advanced searching and multiuser downloads. It is worthwhile for libraries investigating paid aggregator options to spend time learning about the different features available to make the best choice for their users. For example, some paid options offer discounts for bundling certain full-text standards purchases together. Table 5.1 lists common standards aggregators.

Standards Developing Organization Databases

Many SDOs offer free and paid search options to the standards developed by that SDO, and occasionally other SDOs. The SDOs that sell standards are frequently professional societies that also publish journal articles, conference papers, books, and so forth. Like standards aggregators, the free search options typically allow for the purchase of single-user PDFs and/or hard copies of standards, while the subscription platforms provide premium features. Common SDOs that sell individual standards and access to subscription standards platforms are listed in Table 5.2.

TABLE 5.1. *Common Standards Aggregators*

ANSI	Offers a free search option: ANSI Webstore (https://webstore.ansi.org) and a paid subscription option: Standards Connect (https://webstore.ansi .org/Info/StandardsSubscriptions)
IHS	Offers a free search option: IHS Markit Standards Store (https://global .ihs.com), and a paid subscription option: IHS Engineering Workbench (https://ihsmarkit.com/products/engineering-workbench-platform.html [note: now known as Standards Store by Accuris])
MadCad	Offers both free search and paid subscription access options through the MadCad platform (https://www.madcad.com)
Techstreet	Offers a free search option: Techstreet Store (https://www.techstreet.com), and paid subscription option: Techstreet Enterprise (https://discover .techstreet.com/solutions/why-techstreet-enterprise)

Library Catalogs/Discovery Layers

Many academic libraries encourage the use of their library catalog or discovery layer for finding standards and/or access, depending upon their standards holdings and system interoperability. In some cases, libraries purchase hard copy standards, catalog the documents, and add the standards to their print collections so that the physical copies are discoverable and available to users. With digital standards collections, the ability to use library catalogs/discovery layers for standards discovery and access depends upon system interoperability. Standards aggregator databases (e.g., IHS Engineering Workbench, Techstreet) are not currently interoperable with library catalogs/discovery layers.

Additionally, some SDO databases are not currently compatible with library catalogs/discovery layers for searching and discovering standards. System compatibility may depend upon the particular library catalog/discovery layer a library uses and the specific SDO database(s) the library subscribes to. Librarians should test the discovery and access of digital standards collections before promoting the use of the catalog/discovery layer to users.

TABLE 5.2. *Selected SDOs that Offer Full-Text Standards for Purchase*

SDO	Digital Platform	Description of Standards Subscriptions Available
American Society of Civil Engineers (ASCE, https://www.asce.org)	ASCE Library platform (https://ascelibrary.org)	ASCE standards
American Society of Mechanical Engineers (ASME, https://www.asme.org)	ASME Standards Collection (http://asmestandardscollection.org/Login.aspx)	ASME standards
ASTM International (ASTM, https://www.astm.org)	ASTM Compass platform (https://www.astm.org/products-services/enterprise-solutions/astm-compass.html)	ASTM standards, as well as standards from several other organizations, such as the American Association of State Highway and Transportation Officials (AASHTO), the American Concrete Institute (ACI), the American Petroleum Institute (API), the American Welding Society (AWS), International Organization for Standardization (ISO), and Universal Oil Products (UOP) are available on ASTM Compass (as of 12/13/2021)
Institute of Electrical and Electronics Engineers (IEEE, https://www.ieee.org)	IEEE Xplore (https://ieeexplore.ieee.org/Xplore/home.jsp)	IEEE and SMPTE standards (as of 12/13/2021)
SAE International (SAE, https://www.sae.org)	SAE Mobilus (https://saemobilus.sae.org)	SAE standards

Other Library Databases

Outside of standards aggregator and SDO databases, there are other library data-bases that index or provide full-text access to selected standards. Table 5.3 lists additional library databases with standards content (citations only or full text, as noted).

TABLE 5.3. *Selected Library Databases with Standards Content (as of 12/13/2021)*

Library Database	Description of Standards Content
Compendex (an Elsevier Abstracting & Indexing— A&I—database available on the Engineering Village platform)	Currently indexes standards from twelve SDOs, including ASTM and IEEE [2].
Ebook Central (a ProQuest database)	Currently provides full-text access to selected American Water Works Association (AWWA) standards.
INSPEC (an A&I database from the Institute of Engineering and Technology (IET) available on multiple platforms)	Currently indexes selected standards (largely IEEE).
Knovel (an Elsevier database)	Currently indexes and provides options for subscribing to the full text of many standards as part of its subject-area collections. For example, the Knovel subject area "Welding Engineering and Materials Joining" currently includes the full text of AWS B4.0 2016: Standard Methods for Mechanical Testing of Welds (8th edition).
Technology Collection (a ProQuest database)	Currently indexes selected American Institute of Aeronautics and Astronautics (AIAA) standards and provides full-text access to selected National Association of Corrosion Engineers (NACE) standards.

DISCOVERY AND ACCESS CHALLENGES

Locating and accessing standards documents can be a challenging task for both librarians and users. There are obstacles related to the cost of standards, searching for standards, and full-text access restrictions.

Cost of Standards

In general, when deciding what materials to purchase, librarians consider many factors, including the materials budget, cost-per-use, user requests, and how rapidly the material may go out of date. Library material budgets have been challenged for many years, notably with the 2008 global economic crisis [3] and the impact of the COVID-19 pandemic [4]. All the while, the cost of purchasing standards has been an issue for libraries for several decades [1]. Even though, as shown in this book, standards are related to many disciplines, standards collections typically have been held by academic engineering libraries [1]. This may give the impression that standards benefit only a small number of users, especially at academic institutions with small to moderate-sized engineering programs.

 With regard to library users, some instructors do incorporate standards into their courses, especially in engineering and engineering technology programs accredited by ABET [5, 6]; however, in a survey of engineering technology faculty, Khan, Karim, and McLain [7] found that nearly one-third of respondents do not teach about standards in their programs, and almost half reported their own knowledge about standards as insufficient and a barrier to incorporating standards into their teaching. When users request standards, in many cases, it is individual researchers or small groups (e.g., engineering capstone teams) making the requests. While standards teaching practices and user requests vary by program and institution, in libraries that do not get many requests, it can be difficult for librarians to spend funds on standards. This issue is compounded by the fact standards are reviewed and updated regularly (i.e., every five years for ISO standards), and may be regarded as going out of date quickly.

Searching for Standards

Librarians and users encounter multiple challenges when searching for standards. These challenges include unfamiliar terminology, indexing practices, platform interoperability, and identifying the most relevant standards for a particular need.

Some users know the exact standard(s) they need, but others, especially students starting new projects, do not know the particular standard(s) that are relevant and necessary for their work. For example, an engineering student team may be working on a project to design a safer, more accessible chest of drawers for clothing storage. In searching for applicable standards, students may first try the keywords "dresser" and "standards" in Google Scholar, a platform they're very familiar with for academic searching, and not find any relevant results since Google Scholar does not index standards. Next, the students may try searching the keyword "dresser" in either a free or paid standards aggregator database and find the majority of results relate to automotive wheel dressing applications, not clothing storage. They may be able to identify at least one standard that relates to clothing storage, such as ASTM F2057-19, "Standard Safety Specification for Clothing Storage Units," but be at a loss when trying to determine if there are additional standards from other SDOs on the topic.

One solution that could help alleviate this issue is the use of classification searching rather than keyword searching. The International Classification for Standards (ICS) is a hierarchical indexing system for standards and related documents that were developed by ISO [8]. This system is integrated with some standards search engines, such as the subscription aggregator platform IHS Engineering Workbench; however, it is not integrated well or at all with many standards search engines.

Additionally, users may experience confusion and setbacks when attempting to locate the full text of a relevant standard if they discover the document through an aggregator database and their library does not subscribe to the standard via that aggregator database but rather through another channel. For budgetary or other reasons, a library may not be able to subscribe to all the standards its users need through an aggregator database. The library may opt to obtain some standards via aggregator databases subscriptions, some through SDO standards platforms, and some in hard copy format; however, since these systems are not interoperable, users may encounter unexpected paywalls and time delays, when in actuality their library subscribes to the standard they need—just not through the search engine they selected. There is one library database that many engineering libraries subscribe to, Compendex (mentioned previously), that indexes standards and is interoperable with many SDO databases [2].

One further complication with standards searching that librarians and users encounter is determining which standard(s) are the most relevant for a situation, especially when there are several standards applicable to a topic. For example, a few years ago there were seven SDOs working on developing standards related to additive manufacturing [9]. When multiple standards exist on a topic, it can be difficult for libraries to determine the "best" option(s) to purchase with their limited funding. This can be especially problematic when there is little citation information and no full-text preview available.

Full-text Access Restrictions

There are multiple full-text access challenges for both digital and hard copy standards, including digital rights management (DRM) restrictions, document watermarking, and document sharing limitations [1, 10].

DRM restrictions are placed on some standards documents as a way for SDOs to protect their intellectual property and attempt to prevent piracy. In order to open downloaded PDF documents with DRM restrictions (also known as "secure" PDF files), users must install a FileOpen plugin that restricts document access to that machine. This can be problematic for many reasons, including when library users work at library-owned computer terminals and not their own machines. As Cusker noted, there may also be additional DRM impediments with standards, such as printing restrictions and document accessibility time limitations [10].

Another challenge is document watermarking, where SDOs "mark" each page of a standard with details like the customer's name and email address. The practice of watermarking standards can occur with both digital and hard copy purchases. This is particularly concerning if standards are purchased with the intent of being added to the library's collection and contain markings with personal names and contact information. To alleviate this issue, many libraries strive to use purchase methods that are identifiable to the institution, but not individuals [10].

Lastly, traditional document sharing through interlibrary loan is difficult, if not prohibited, with standards documents [1, 10]. The license agreements put into place by SDOs frequently contain language that prevents loaning the documents. Additionally, even in cases where libraries can legally lend standards (e.g., some hard copy purchases), librarians may still be hesitant to lend the material, given uncertainty about the license agreement restrictions and/or concerns about the standard not being returned.

FREELY AVAILABLE FULL-TEXT STANDARDS

Several SDOs and other standards-related organizations make their standards available for free, openly, or for educational purposes. Table 5.4 lists selected organizations and programs that currently offer standards for free in some capacity.

TABLE 5.4. *SDOs/Standards Programs with Free Full-Text Standards Access (as of 12/13/2021)*

SDO/Standards Program	Description of Free Access
Air-Conditioning, Heating and Refrigeration Institute (AHRI)	All AHRI standards are freely available for anyone to download. https://www.ahrinet.org/standards
American National Standards Institute (ANSI) University Outreach Program	Upon request, ANSI provides free access to selected International Electrotechnical Commission (IEC) and ISO standards for faculty members and students for classroom use. This program cannot be used to fulfill standards requests for research purposes. A faculty member must make the request from a dot edu email address, and the turnaround time is approximately one week. https://www.ansi.org/education /activities/standards-university-outreach Additionally, through this ANSI program, the American Dental Association (ADA) makes selected standards available to dental faculty members and students. See https:// www.ada.org /resources/practice/dental-standards/university-outreach
ANSI Incorporated by Reference (IBR) Portal	Selected standards and codes from many different SDOs that are referenced in the U.S. Code of Federal of Regulations are freely available through this portal. https://ibr.ansi.org.
ASTM International (ASTM)	Users who register for an account can view ASTM standards that are incorporated into U.S. regulations freely online in the ASTM Reading Room. Documents cannot be downloaded or printed. https://www.astm.org/products-services/ reading-room.html

TABLE 5.4. *Continued*

American Society of Heating, Refrigerating and Air-Conditioning Engineers (ASHRAE)	Users can view ASHRAE standards and guidelines freely online. Documents cannot be downloaded or printed. https://www .ashrae.org/technical-resources/standards-and-guidelines
Ecma International	All Ecma standards are freely available for anyone to download. https://www.ecma-international.org/publications-and-standards /standards
European Telecommunications Standards Institute (ETSI)	All ETSI standards are freely available for anyone to download. https://www.etsi.org/standards#Pre-defined%20 Collections
Industrial Truck Standards Foundation (ITSDF)	Users who register for a free temporary account can access ITSDF B56 standards freely. http://www.itsdf.org/cue/b56 -standards.html
ISO/IEC Joint Technical Committee (JTC) 1 Information Technology	Selected ISO/IEC JTC 1 standards are freely available for anyone to download. https://standards.iso.org/ittf /PubliclyAvailableStandards/index.html
International Telecommunications Union (ITU)	Most ITU standards are freely available for anyone to download. ITU-D Recs: https://www.itu.int/rec/D-REC/en; ITU-R Recs: https://www.itu.int/pub/R-REC; ITU-T Recs: https://www.itu .int/en/ITU-T/publications/Pages/recs.aspx
Military Standards and Specifications	Most military standards (MIL-STD) and specifications (MIL-SPEC), as well as other defense handbooks and guidance documents, are freely available for anyone to download on EverySpec (http://everyspec.com) and ASSIST QuickSearch (https://quicksearch.dla.mil/qsSearch.aspx).

TABLE 5.4. *Continued*

National Fire Protection Association (NFPA)	Users who register for an account can view standards freely online with NFPA's Free Access view. Documents cannot be downloaded or printed with NFPA Free Access. https://www.nfpa.org/Codes-and-Standards/All-Codes-and-Standards/Free-access
National Information Standards Organization (NISO)	All NISO standards are freely available for anyone to download. https://www.niso.org/publications
National Operating Committee on Standards for Athletic Equipment (NOCSAE)	All NOCSAE standards are freely available for anyone to download. https://nocsae.org/standards/standards-matrix/#/all/performance/current
Simulation Interoperability Standards Organization (SISO)	All SISO standards are freely available for anyone to download https://www.sisostds.org/ProductsPublications/Standards.aspx
Sporting Arms and Ammunition Manufacturers' Institute (SAAMI)	All SAAMI standards are freely available for anyone to download. https://saami.org/technical-information/ansi-saami-standards
Snell Foundation	All Snell Foundation standards are freely available for anyone to download. https://smf.org/stds
Underwriters Laboratories (UL)	Users who register for an account can view standards freely online with UL's Free Digital View. Documents cannot be downloaded or printed with Digital View. https://www.shopulstandards.com/Catalog.aspx. For more information, see https://www.ul.com/news/qa-complimentary-online-access-ul-standards

Additionally, in response to the COVID-19 pandemic, many SDOs offered selected standards freely that relate to areas in need of rapid action due to the pandemic, such as medical devices, personal protective equipment, global health, health technology, and business continuity planning [11]. This instance demonstrated that in a time of a global health crisis, SDOs recognized the importance of access to standards in a time of rapid response. SDOs made standards available in read-only format and some required users to create an account to access the documents.

SUMMARY

Discovering and accessing standards can be more challenging than traditional academic library resources. There are many different types of standards search engines (freely available and for pay) that index and provide full-text options for hard copy and digital standards purchasing; however, most are not interoperable. Standards discovery is complicated by the lack of widespread use of the International Classification for Standards (ICS) system and the absence of standards indexing in Google Scholar. Additionally, the cost of standards and SDO document security practices, such as implementing DRM restrictions, make full-text access to standards difficult for libraries. Fortunately, there are many SDOs that make their standards available freely, either publicly or for educational purposes.

REFERENCES

1   M. Phillips, "Standards collections: Considerations for the future," *Collection Management, 44(2–4),* Jul. 2019, pp. 334–347. https://doi.org/10.1080/01462679.2018.156 2396

2   M. Phillips, "Technical standards in Compendex," *Issues in Science and Technology Librarianship, (99),* 2022. https://doi.org/10.29173/istl2621

3   M. Savova and J. S. Price, "Redesigning the Academic library materials budget for the digital age: Applying the power of faceted classification to acquisitions fund management," *Library Resources & Technical Services, 63(2),* Apr. 2019. https://doi.org/10.5860/lrts.63n2.131

4   "Academic library strategy and budgeting during the COVID-19 pandemic," *Ithaka S+R.* https://sr.ithaka.org/publications/academic-library-strategy-and-budgeting-during-the-covid-19-pandemic (accessed Dec. 13, 2021).

5   P. McPherson, M. Phillips, and K. Reiter, "Integrating technical standards into ET curricula to meet ABET standards and industry needs," presented at the ASEE

2019 Conference for Industry and Education Collaboration (CIEC), Feb. 2019. https://peer.asee.org/31507 (accessed Dec. 14, 2021).

6 D. Solomon, Y.-T. T. Liao, and J. T. Chapin, "Maximizing the effectiveness of one-time standards instruction sessions with formative assessment," presented at the 2019 ASEE Annual Conference and Exposition, Jun. 2019. https://doi.org /10.18260/1-2--33087

7 A. S. Khan, A. Karim, and J. A. McClain, "The state of the use of standards in engineering and technology education," Jun. 2013. https://peer.asee.org /the-state-of-the-use-of-standards-in-engineering-and-technology-education (accessed Dec. 14, 2021).

8 ISO, *International Classification for Standards (ICS)*, 7th ed., 2015. https://www.iso .org/publication/PUB100033.html (accessed Dec. 14, 2021).

9 P. Witherell and Y. Lu, "Building connections through standards landscaping," *The Journal of SES—The Society for Standards Professionals, 69(3), pp. 1–8.*

10 J. Cusker, "Adding Individual technical standards to a library collection: A case study and a proposed workflow," *Collection Management, 45(2), Apr. 2020, pp. 124– 138. https://doi.org/10.1080/01462679.2019.1650864*

11 American National Standards Institute, https://www.ansi.org/news/standards -news/2020/04/standards-organizations-open-access-to-standards-and-resources -to-support-covid19-response-06 (accessed Dec. 13, 2021).

6

Standards Collection Development

Erin M. Rowley, University at Buffalo (SUNY)

INTRODUCTION

This chapter is intended for librarians who are responsible for or interested in standards collection development. It is recommended that faculty interested in procuring standards contact the appropriate librarian at their own institution, or refer to Chapter 5, "Discovering and Accessing Standards," for more information.

POLICY DEVELOPMENT

Per the Wetzel et al. study that conducted a survey of the Association of Research Libraries (ARL) on accessing standards, the majority of survey respondents indicated they do not currently have a specific policy in terms of the collection of standards [1]. In addition, many libraries do not have specific funds set aside for standards purchasing, despite standards being purchased at those institutions.

As of December 2021, conducting a search of websites ending in ".edu" for "library collection development policy," and then searching resulting documents for "standards" did not yield any results related to technical standards as defined by this publication.

In terms of resources to develop a collections policy related to standards, very little exists in books devoted to library collection development practices. Many books related to collection development in libraries, more generally, and in academic libraries, specifically, were consulted for this chapter; however, none addressed the topic of standards, industry standards, or technical standards used in engineering or other academic areas [2–14]. Perhaps the best resource developed for librarians that discussed standards in a library collection is the 2013 book, *Engineering Libraries: Building Collections and Delivering Services* [2]. Thompson's chapter, "Grey Literature in Engineering," which is the same as his article of the same title, very succinctly states:

> In the past, it has been difficult to provide standards access to researchers and practicing engineers. Onsite paper format collections were expensive to subscribe to, time-consuming to update, and almost impossible to catalog efficiently [15].

Thompson's chapter does not cover standards collection policies, but rather access to such documents. It nonetheless illustrates, to current and potential engineering librarians, the challenges of acquiring standards, let alone creating a formal collection development policy for them.

Books that address the importance of standards overall typically did not refer to libraries at all. The few that did mention policies or practices for acquiring standards were done so from the perspective of a private company or something similar [16–19]; however, there is still some information to be gleaned from those cases. As Batik states, companies should be prepared to budget accordingly due to the continual cost of building and maintaining a standards collection [16]. Crawford spoke of dedicating physical space to the standards purchased. Of course, as discussed in Chapter 5, many who utilize standards choose to access them electronically; the Crawford book was published in 1983, which certainly explains why electronic access was not mentioned [17].

Batik also pointed out that it is helpful to have a designated person manage a standards collection, thereby demonstrating the specialization one possesses when they are knowledgeable about standards as an information source and in the process of procuring said documents.

As books related to library collection development do not tackle the topic of standards, and books related to standards specifically do not touch on the topic of libraries, collection policy information related to standards is most often found in journal articles and conference papers.

MODELS OF COLLECTION PRACTICES

In terms of models for the collection of standards, there are several that exist based on current literature and anecdotal evidence shared among engineering librarians. However, many libraries do not seem to follow one model exclusively, based on the author's experience. Each of the most common models will be discussed individually below. Often libraries just beginning the process of obtaining standards will start with one model and evolve to a new model entirely or add an additional model to their current practices [20].

Purchase in Print

Purchasing standards in print was, like for almost any library item, the original and only way to gain access to these documents before electronic access was established. However, since so many different standards developing organizations (SDOs) exist, there have been and continue to be different ways to go about this.

ASTM International, for example, is one SDO that produces an extensive print series of books annually for their standards. At the time of publication, a complete annual set of ASTM standards would be 85 volumes [21]. Another example is the American Society of Mechanical Engineers (ASME) Boiler and Pressure Vessel Code, published annually. This comes in a 32-volume set [22]. Of course, there are several concerns to keep in mind when purchasing standards in print, especially with a multivolume set. First, space concerns are continually an issue for libraries everywhere, regardless of type, size, or collection [23–30]. Standards are frequently shelved in reference areas of the library so that they are readily accessible, but regardless of location, these sets take up a vast amount of shelf space. Also, it is important to note that all physical items need to be cataloged and processed before they are available to users. In the case of purchasing a single print standard, the document will likely not come in a format that is conducive to being placed on a library shelf, meaning the standard may need some sort of binding, which takes time and money, before being placed in the open stacks.

Conversely, Cusker explores a model of adding individual technical standards to a library collection in his case study-based article [23]. Dunn and Xie explored adding individual standards to the collection at their library through a single specific vendor [24]. In both of these papers, a specific workflow for purchasing the standards, communicating with the patron, and delivering the standards was detailed.

Purchasing Electronically: Subscriptions

With an increasing percentage of libraries' collections being purchased electronically [31], standards are no different. Standards publishers and standards aggregators have been making the move to offer standards electronically to customers for many years. In some cases, this means purchasing standards individually as PDF downloads, which can be locked to one single computer station (this will be discussed more in the next subsection); however, there are options for libraries and librarians to obtain electronic access to many different types of standards at an institutional level, similar to other subscription databases for journals. The American Society of Civil Engineers (ASCE), the American Society of Mechanical Engineers (ASME), ASTM International, the Institute of Electrical and Electronics Engineers (IEEE), and SAE International are all SDOs, to name only a few, that now offer electronic standards access through an annual subscription via an online portal.

Other SDOs not listed above may still allow for institutional-level electronic access but may rely on a standards aggregator database to provide that access. See the section below entitled "Aggregator Databases and Services" for more details.

One benefit of electronically purchasing standards, which are available through an online portal for access, is allowing for access anywhere. Just as students can now search for journal articles from wherever they are located (typically with their university login credentials for verification purposes), users can now search for and access standards in a similar manner.

Another benefit to this model of access is currency. By purchasing in print, standards can, in some cases, quickly become outdated or canceled; however, in almost all cases, a subscription to an SDO online portal means that users will always have access to the most current version of the standards. In some cases, some standards portals like ASTM Compass also provide access to all previous versions of standards as well, which can be useful.

Lastly, electronic access to standards helps to address the common library issue of space. Print standards volumes can often take up rows and rows of shelf space. Even purchasing individual standards in print will require some physical space for the standards to be stored. Purchasing individual print standards, as mentioned previously, also requires time and effort to catalog and process the item before making it available to users.

Purchasing Electronically: Individual PDFs

Another model of collection development for standards is purchasing electronically, but instead of institutional-wide access, standards are purchased individually as a single PDF. Some libraries have implemented this model to obtain standards for students on-demand, where one student may need a standard for a senior capstone design course, but it is unique to that student's needs and therefore the need for continual, institutional access is unnecessary [28]. For libraries implementing this model, patrons may be asked to fill out an electronic form on the library website, or to contact the librarian directly, requesting the specific document they need. When the standard is purchased, a PDF document of the standard is then delivered directly to the user, often via email.

Some libraries that utilize this model have a set budget for one-off standards purchases each fiscal year. When the budget has been used, standards purchases are then cut off until the next fiscal year [23–24]. In other cases, there is no set budget for these types of purchases and librarians may include the purchase of these specialized documents with the line item for other one-time purchases like books and e-books.

Aggregator Databases and Services

Purchasing individual standards or sets of standards in an electronic format can often (but not always) be done directly from the SDO website; however, there are many different standards aggregator databases or services that exist for customers to purchase standards from multiple SDOs from a single website. In addition to acting as a point of purchase for standards from many different organizations, in most cases these databases/websites allow users to search the catalog of available standards by keyword. This is extremely helpful for a variety of users who want to see if one or more standards exist for a given product or process, including but not limited to professionals, students, faculty, and librarians.

It is important to note at the outset that access to standards via standards aggregators is not consistent. The ability to purchase electronic access, especially at an institutional level for academic libraries, is dependent upon the standards developing organization. Some SDOs will allow for electronic access to individual standards, where access is purchased for each specific standard needed. Other SDOs will only allow electronic access to their standards if the entirety of the standards catalog is purchased—and typically for a high price tag.

The most popular aggregators in the United States (although they can be used internationally as well) include Techstreet, IHS Global, the ANSI Webstore, and SAI Global. Each of these tools will be briefly described below and include some basic information regarding standards coverage.

Techstreet

Techstreet is a popular standards aggregator database with more than 130,000 users from academic, corporate, and government entities. It is described as "part of Global Knowledge Solutions, a for-profit subsidiary of the American Society of Mechanical Engineers (ASME)" [32].

It boasts access to more than 150 standards publishers with more than 550,000 standards and codes available for purchase. Techstreet offers the purchase of standards in both print and electronic formats. They also offer a free service to track standards updates, so anyone can be alerted if a new version of a standard is published.

Techstreet also offers electronic access to standards via the Techstreet Enterprise platform. Individual standards or groups of standards can be selected to access for an annual subscription fee. New standards can be added to the access at any time for a prorated amount. Price quotes for new standards access are available directly through the Techstreet Enterprise platform with a single click. The Techstreet Enterprise platform can be used by academic libraries (and other organizations) to give all users immediate electronic access to the standards they subscribe to.

IHS Global Standards Store

IHS Global offers a "broad base of engineering data from research and design to manufacturing and repair" [33]. Their standards store includes standards and specifications from more than 460 technical societies globally. Like Techstreet, they also offer a variety of purchase and delivery options, as well as free update services to stay up-to-date on changes made to specifically selected standards.

Additionally, IHS Global also allows for custom online standards collections where you select only the documents you need. Access is provided through the IHS Engineering Workbench [34].

ANSI Webstore

ANSI, or the American National Standards Institute, is an SDO in the United States and the official U.S. representative to the International Organization for Standardization (ISO). In addition, ANSI offers the ANSI Webstore where customers can search for and purchase standards. Standards, like Techstreet and IHS Global, can be purchased in electronic (PDF) or print form. The ANSI Webstore offers over 200,000 standards from more than 150 standards publishers [35].

Similar to the other aggregators mentioned, ANSI offers "Standards Connect" to provide online access to select standards for multiple concurrent users. ANSI also provides the option of "standards packages" where a collection of standards can be provided at a discounted rate; however, standards packages are limited to one user per license and focus on downloading standards rather than online access [36].

SAI Global

SAI Global provides over 1.6 million standards from more than 360 publishers. Standards can be searched and purchased individually—in print or PDF electronic format—through the Infostore [37]. In addition, as with others mentioned previously, SAI Global offers a solution to access and manage standards electronically through the i2i platform [38]. The i2i platform also provides reporting functionality and dashboards to determine the most used, accessed, and requested standards to optimize your subscription.

STORAGE AND ORGANIZATION

The collection of standards generally falls into two large buckets: print form and electronic form. Organization and storage depend largely on which form the standard was obtained in, therefore, each will be addressed separately below.

Storage and Organization of Standards in Print Form

Standards purchased in print form typically come as an individual document, if one single standard is purchased separately, or they can be purchased as a set that comes in a large volume or volume set (e.g., ASTM standards series published annually as a multivolume set). Libraries have multiple options when purchasing standards individually. One popular option is to purchase the standard from a standards aggregator database such as those mentioned in the previous section [25, 28]. These vendors allow users to purchase standards individually, typically in print or electronic (PDF) format (or, in some cases, an option to purchase both formats simultaneously also exists). When purchasing in print format, these individual standards are mailed to the purchaser.

Another option for libraries looking to purchase standards individually in print form is to purchase directly from the SDO, such as ISO, ASTM, ASME, ASCE, and so forth. Again, there typically is an option to also purchase electronically. Still, these purchases are usually "locked" to a single computer, preventing the user from sharing the PDF of the standard with others via email or a shared drive. Depending on the standard needed, there may be other places where a standard can be purchased from in print. Some standards, for instance, are available from library acquisition websites such as GOBI (offered by Yankee Book Peddler, Inc.) or OASIS (offered by ProQuest). If a library needs to purchase a standard in print, it is best to consult several different options to ensure the best price, although most librarians will find the price is fairly consistent from one provider to the next as it is set by the SDO that published the standard.

When purchasing standards in print, especially when purchasing single standards, it is important to consider several things. First, cataloging standards can take more time in comparison to books. Copy cataloging is not often an option for standards, and some catalogers may not be overly familiar with cataloging standards. Second, individual standards are not usually delivered as a traditionally library-bound item, so additional time and money may need to be spent to place the standard in a cover to protect it and ensure longevity. Last, many libraries will elect to shelve standards in their reference collections to ensure they can be accessed easily by patrons, but also to prevent theft [29]. Standards can be costly documents, as mentioned previously, especially since they are not overly lengthy (in most cases) and can be updated frequently. Libraries may choose to recatalog superseded versions of standards or cancelled standards in the general collection and only shelve current print standards in reference [29].

Purchasing standards in print has both benefits and disadvantages. By purchasing in print, it is usually just a one-time cost, which is often the most manageable by libraries (in comparison to continuing commitments like subscriptions). Of course, if standards are updated frequently, this can become a recurring cost for libraries, so this is something to keep in mind. Print access also can allow, under some circumstances, the document to be loaned via interlibrary loan; however, librarians are cautioned to read terms and conditions when purchasing individual standards very carefully, as interlibrary loans may not be allowed by certain SDOs.

The disadvantages of purchasing in print, unfortunately, can outnumber advantages. When purchasing any item in print, one must consider the time it takes to receive the item, shipping charges, as well as the cost and time to catalog and process the item. In addition, access to the standard is dependent on when the library is open. It also requires patrons, in most cases, to come to the library to access the standard, especially if it is only available in a reference collection. These factors should all be considered when obtaining standards for a library collection.

Storage and Organization of Standards in Electronic Form

For standards obtained electronically, many libraries will access these standards via a database or other electronic platform. Electronic access is dependent upon if the SDO allows for institutional access. Standards aggregators and individual standards developing organizations will often allow for a one-time electronic download of a standard in PDF form; however, as mentioned previously, this type of purchase typically is locked to a single computer station, which is not ideal for a library. Even if the standard could be made available on a single public computing station, it would severely limit access, ultimately defeating the purpose of buying electronically. There are some cases of libraries reporting purchasing standards electronically as a one-time PDF download and sending them directly to the patron who requested the standard [23, 24]. This method is not used at all libraries, though, and could create the need for duplicate purchases if more than one patron requests the same standard.

The option that is becoming increasingly popular with academic libraries is to purchase access to a set of standards and access them through a website or other electronic platform [24, 25, 28]. For example, many academic libraries with engineering programs receive access to all IEEE standards via the IEEE Xplore database, providing access to IEEE journal articles and conference papers. Another common platform for academic libraries to subscribe to where engineering programs exist is the ASTM Compass database, which provides access to ASTM standards. Other platforms that provide access to standards include SAE Mobilus from SAE International and ASCE Research Library from the American Society of Civil Engineers.

As mentioned previously, aggregators also provide electronic access to standards. Access to these standards is dictated by the individual SDO. For example, ISO standards must be requested and subscribed to individually; however, "blocks" of standards can be purchased from an aggregator like Techstreet to access British Standards Institute (BSI) standards. A predetermined number of standards, for example, 10, make up the "block" and are prepaid. Then the librarian can select which standards to obtain access to, typically based upon patron request. These can be changed annually when the block of standards is prepaid again. In yet another example, UL standards from the Underwriters Laboratory can be purchased from standards aggregators, but must be purchased completely—in other words, it's all or nothing in terms of access, which is reflected in the cost.

Access to standards electronically, unless being purchased as a one-off, PDF download, is treated as a subscription in that it is an annual cost; however, in almost all cases, this access allows users to download standards, which is helpful if they need to be printed or accessed at a later time. In some cases, such as with ASTM Compass, access to the database of standards allows the user to view the current version of the standard as well as older, superseded versions. ASTM Compass also provides access to redline versions that compare the current version to the previous version, illustrating where changes have occurred.

SUMMARY

Crawford perhaps said it best in his 1986 book, "libraries must fight for the standards they want" [17]. Standards are typically viewed as a "high cost" item for libraries, especially when considering the length of the documents and how often they are reviewed and updated. In addition, when purchasing standards in print, the labor costs must also be considered for obtaining the documents, cataloging the standard, and, in some cases, providing special binding. Due to the limitations of working with physical documents, many libraries have moved toward electronic access via one or several of the databases listed above; however, this also translates to a higher cost as access is now a continuing commitment instead of a one-time cost. In a time when library budgets stay flat or are cut each year, adding continuing commitment costs typically means canceling something else. It can also require justification of the purchase, which takes the librarian(s) involved additional time and effort. Faculty statements requesting the resources also may be needed to move forward with a new database subscription; however, costs aside, engineering librarians, STEM librarians, and other librarians working with standards continue to fight for access to these technical documents.

REFERENCES

1 D. A. Wetzel and K. Grove, "Accessing engineering standards: A study in ARL best practices for acquiring and disseminating standards," presented at the 2021 ASEE Virtual Annual Conference Content Access, Virtual Conference, July 2021.

2 T. W. Conkling and L. R. Musser, *Engineering libraries: Building collections and delivering services.* New York: Routledge, 2013.

3 W. Disher, *Crash course in collection development.* Westport, CT: Libraries Unlimited, 2007.

4 M. Fieldhouse and A. Marshall, *Collection development in the digital age.* London: Facet Publishing, 2012.

5 E. Futas, ed., *Collection development policies and procedures.* Phoenix, AZ: Oryx Press, 1995.

6 J. T. Gillespie and R. J. Folcarelli, *Guides to library collection development.* Englewood, CO: Libraries Unlimited, 1994.

7 V. L. Gregory, *Collection development and management for 21st century library collections: An introduction.* New York: Neal-Schuman Publishers, 2011.

8 F. W. Hoffmann and R. J. Wood, *Library collection development policies: Academic, public, and special libraries.* Lanham, MD: Scarecrow Press, 2005.

9 S. Holder, *Library collection development for professional programs: Trends and best practices.* Hershey, PA: IGI Global, 2012.

10 P. Johnson, *Fundamentals of collection development and management.* Chicago: ALA Editions, 2013.

11 P. Johnson, *Fundamentals of collection development and management.* Chicago: ALA Editions, 2014.

12 A. M. Morrison, Managing electronic government information in libraries: Issues and practices. Chicago: American Library Association, 2008.

13 M. Pearce, *Non-standard collection management*. Aldershot, England: Ashgate, 1992.

14 R. J. Wood and F. W. Hoffmann, *Library collection development policies: A reference and writers' handbook*. Lanham, MD: Scarecrow Press, 1996.

15 L. A. Thompson, "Grey literature in engineering," *Science & Technology Libraries*, 19(3-4), 2001, pp. 57–73. https://doi.org/10.1300/J122v19n03_05.

16 A. L. Batik, *The engineering standard: A most useful tool*. Ashland, OH: Book Master/El Rancho, 1992.

17 W. Crawford, *Technical standards: An introduction for librarians*. White Plains, NY: Knowledge Industry Publications, 1986.

18 S. M. Spivak and F. C. Brenner, *Standardization essentials: Principles and practice*. New York: Marcel Dekker, 2001.

19 C. D. Sullivan, *Standards and standardization: Basic principles and applications*. New York: M. Dekker, 1983.

20 L. J. Pellack, "Industry standards in ARL libraries: Electronic and on-demand," *Collection Building*, 24(1), 2005, pp. 20–28.

21 ASTM International, "Complete set [annual book of ASTM standards]." https://www.astm.org/products-services/astm-bos-all.html (accessed August 2, 2021).

22 The American Society of Mechanical Engineers (ASME), "Boiler and Pressure Vessel Code 2021 Complete Set." https://www.asme.org/codes-standards/find-codes-standards/bpvc-complete-code-boiler-pressure-vessel-code-complete-set/2021/print-book (accessed August 2, 2021).

23 J. Cusker, "Adding individual technical standards to a library collection: A case study and a proposed workflow," *Collection Management*, 45(2), 2019, pp. 1–15. https://doi.org/10.1080/01462679.2019.1650864.

24 L. K. Dunn and X. Shiyi, "Standards collection development and management in an academic library: A case study at The University of Western Ontario Libraries," *Issues in Science & Technology Librarianship*, 87, 2017, https://doi.org/10.5062/F4KK9928.

25 K. A. Kozak, "Standards, standards, where might you be?," *ASEE North Midwest Section Conference*, Iowa City, Iowa, 2014, pp. 1–8. https://doi.org/10.17077/aseenmw2014.1039.

26 L. R. Musser, "Standards collections for academic libraries," *Science & Technology Libraries*, 10, 1990, pp. 59–71. https://doi.org/10.1300/J122v10n03_05.

27 J. Papin-Ramcharan, A. Dolland, and R. A. Dawe, "Making engineering standards available at the University of the West Indies: Perspectives of a developing country," *Collection Building*, 30(2), 2011, pp. 86–93. https://doi.org/10.1108/01604951111127452.

28 M. Phillips, "Standards collections: Considerations for the future," *Collection Management, 44*(2–4), 2019, pp. 334–347. https://doi.org/10.1080/01462679.2018.1562396.

29 D. Taylor, "Standards collection development in an academic library," *Collection Building*, vol. 18, no. 4, pp. 148-152, 1999. https://doi.org/10.1108/01604959910303280.

30 S. B. Wainscott and R. J. Zwiercan, "Improving access to standards," *2020 ASEE Virtual Annual Conference Content Access*, 2020. https://peer.asee.org/34790 (accessed November 27, 2020).

31 J. K. Frederick and C. Wolff-Eisenberg, "Ithaka S+R US Library Survey 2019," April 2 2020. https://doi.org/10.18665/sr.312977 (accessed November 27, 2020).

32 Techstreet, "Techstreet: Connecting the world to standards." https://discover .techstreet.com/about-techstreet (accessed August 2, 2021).

33 IHS Global, "About us—IHS Global." https://global.ihs.com/about.cfm?&rid =Z56&mid=STDS (accessed August 2, 2021).

34 IHS Global, "Custom online standards collections." https://global.ihs.com/custom _online_collections.cfm?&rid=Z56&mid=STDS&input_doc_number=&input _doc_title= (accessed August 2, 2021).

35 American National Standards Institute (ANSI), "About ANSI Webstore." https:// webstore.ansi.org/Info/About (accessed August 2, 2021).

36 American National Standards Institute (ANSI), "ANSI Webstore." https://webstore .ansi.org (accessed August 2, 2021).

37 SAI Global, "SAI Global Infostore." https://infostore.saiglobal.com/en-us (accessed August 2, 2021).

38 SAI Global, "SAI Global i2i—Information to Intelligence." https://infostore .saiglobal.com/en-us/standards_management (accessed August 2, 2021).

PART III

Standards Curriculum Integration and Requirements

7

Standards Teaching and Learning

Chelsea Leachman, Washington State University, and Daniela Solomon,
Case Western Reserve University

INTRODUCTION

Many courses and programs are continually strapped for time in the classroom, and presenting a new topic can be difficult to either fit into classroom time or program. This chapter includes ways to incorporate technical standards informational literacy into the curriculum, tips for advocating for technical standards instruction, provides instructional assessment activities, and an instruction toolkit. This chapter focuses on advocating for the inclusion of technical standards within the curriculum, whereas specific discipline requirements can be found in Chapters 9–13. This chapter focuses on advocating for the inclusion of technical standards within the curriculum, whereas specific discipline requirements can be found in Chapters 9–13, and case study examples can be found in Part IV.

Despite curriculum integration being considered the most effective way to introduce standards to students [1–5], it is still not a common practice [6]. Integrating technical standards into the curriculum is hindered by several factors [6]. One factor is that curricula for engineering courses are heavy with technical subjects for the majority of a student's academic career, which leaves little room for other topics such as ethics or technical standards [7]. Additionally, the development of new courses and curricular changes are challenging processes, faculty and instructors have little or no knowledge of practitioner standards and, as of the time of this publication, there are no current technical standards textbooks or handbooks. Finally, limited access to standards due to their high cost contributes to a lack of effort in teaching technical standards in academia.

While technical standards have been integrated into engineering curricula during their capstone and senior design courses, recent literature found that standards education in upper division undergraduate courses alone may not be as effective as previously thought. For students to successfully use technical standards, they need to become acquainted with standards earlier in their academic career [8]. Students should learn about standards through many team-based experiences before their final year [9]. Regarding the other disciplines discussed in this book, standards education has not traditionally been integrated into business education [10], health sciences, or law.

Some examples of integrated technical standards into curricula include incorporation of standards into class syllabi, use of standards in other design classes [4, 13–14], development of learning objects [15–16], or the development of standardization courses [17–18]. In addition to curriculum integration, other common practices for standards education are one-shot library instruction sessions on standards [19–20] or campus-wide educational events [7]. Co-op, experiential learning, and internship experiences also offer students opportunities to learn about standards [21–22].

Possible solutions to alleviate the lack of standards information literacy instruction include "Train-the-Trainer" educational initiatives, making technical standards library collections more affordable, developing teaching materials that focus on the fundamentals of standards, providing technical standards literacy modules, and building an understanding of standards applicability to product design, manufacture, and quality control.

TECHNICAL STANDARDS INFORMATION LITERACY INSTRUCTIONAL PLANNING

When preparing to teach a class or series of classes on technical standards, incorporating instruction planning can help meet the class, course, or program goals. Through instructional planning, instructors, faculty, and librarians can focus on incorporating information literacy outcomes into the class or course rather than demonstrating tools or skill-based tasks. Depending on the program, students receive information literacy instruction throughout their academic careers, at different times of need, and in different delivery modes. Many students begin their academic careers learning information literacy skills through writing academic papers and, as they move through the curriculum, become more familiar with their field's information needs and sources. As with all information literacy instruction, technical standards information literacy should be planned to be incorporated at a time of need through course instruction, consultations, or online resources.

Information literacy planning using backward instruction planning can help by focusing on the learning outcomes when developing technical standards interventions. Backward planning is student-centered and "helps students connect theory to practice, reflect on their learning, and construct new knowledge as they build upon prior knowledge and experience" by starting with the learning outcome [23]. The benefits of backward planning include that it can be easily applied to various instruction situations, helps allocate appropriate time to instructional topics, and shifts the focus to the learning activities that will achieve the learning outcomes [23]. The backward planning can be used for one-shot, online, or longer-term instructional scenarios. Incorporating instructional design into teaching technical standards also helps to communicate to internal and external stakeholders the value of instruction.

Backward planning is a three-step process: understanding the program level learning outcomes, crafting specific learning outcomes for the instructional session, and planning learning activities to meet short- and long-term information literacy learning outcomes [23]. The first step involves identifying the long-term goals for the students in a course, such as "They should be able to use discipline-specific technical standards." The second step in the process is breaking the long-term learning outcome into small achievable learning outcomes, such as "Students will be able to use online databases to identify and locate technical standards." Once the outcomes are identified for the specific class session, series of classes, or course planning, instruction moves into thinking about learning activities. In addition to planning the learning outcomes for the instructional session, finding the appropriate timing of technical standards information literacy instruction can be challenging.

When looking for when to introduce technical standards into the program or course, scaffolding can help librarians or instructors focus on introducing technical standards when and where there is a time of need. When scaffolding information literacy instruction, the librarian or instructor can plan out when new concepts are introduced and when previously taught skills can be reinforced. It is recognized that while students might not remember all previously taught information literacy skills, context for the information and reinforcement of these skills should be sufficient. Additionally, the instructor or librarian can provide foundational information literacy skill resources for students to reference outside instruction or consultation.

If you are interested in scaffolding information literacy, a helpful tool when planning out the level and timing of instruction is curriculum maps. Curriculum mapping has been done at the K–12 level and all levels of higher education. Curriculum mapping the current information literacy instruction can give instructors, faculty, or librarians the ability to "examine the [information literacy] curriculum in its entirety" [24]. The process of mapping information literacy instruction allows for identifying gaps, redundancies, and misalignments in the instructional program [25]. This broad view of information literacy instruction can help identify where information literacy instruction can be changed, added, or even removed. Curriculum mapping information literacy can help with student engagement in the classroom. Salisbury and Sheridan found that students experience frustration when skills instruction is repeated in different courses or when skills are presented out of order of difficulty [26].

Curriculum mapping aims to identify who is doing what, how the work is aligned with the goals, and if you are working efficiently and effectively [25]. The curriculum mapping process involves charting instruction throughout the discipline's curriculum to identify courses currently receiving information literacy instruction, strategically identify courses to target for future instructional efforts, and map all required courses and possible instructional levels for the discipline. To populate the curriculum map, information can be collected from the following sources: course catalogs, program data, course syllabi, or library instruction data [27]. The curriculum mapping process allows instructors, faculty, or librarians to strategically and intentionally identify appropriate information literacy access points as it "allows participants to articulate their intended outcome clearly and visually evaluate how those outcomes fit into the student experience" [24].

Different instructional methods should also be considered when scaffolding the curriculum based on time allotment and course outcomes.

Face-to-Face: Instruction is any in-person instruction in a classroom or lab setting. Face-to-Face instruction allows students, the instructor, faculty, or librarian to move around the room for group work or to check on students engaging with the material. Face-to-face also allows for print materials to be shared with the students.

Online Synchronous: Learning happens with students in a live virtual environment. Synchronous online learning environments might also include group work through breakout rooms, as the software allows.

Online Asynchronous: Online learning happens when students interact with the learning object on their own time, but it is a required assignment from an instructor or faculty. Asynchronous learning might include recorded lectures or videos with included quizzes or other question and answer areas.

Passive: Online learning tools would be tools that are not assigned during a course but are available to students in a time of need for instruction. Examples would be LibGuides, online tutorials, websites, or vendor tools that students might use to learn about specific organizations. Other software tools might be available depending on the institution; some examples include Libwizard, Qualtrics, etc.

An example of curriculum mapping from electrical engineering is included and shows the courses that were already receiving information literacy and identified a course as a target for instruction (see Table 7.1). As noted on the curriculum map, the electrical engineering students receive information literacy from an undergraduate services librarian in their first year of English and history courses. Working with the undergraduate services librarians, the skills taught during those courses were identified and mapped to learning outcomes. Students in electrical engineering also take an introductory engineering course in their first year, and during this course, the students receive information literacy support through a LibGuide. The LibGuide was chosen in this case because most students are already receiving information literacy instruction face-to-face in two of their other courses, and most of the skills taught in those courses are transferable to this course. Following the progression of courses, electrical engineering students spend a large part of their second and third years taking technical science and engineering courses, which do not usually incorporate a component of information literacy. By the end of the third year and into their fourth year, the electrical engineering students take courses in professional ethics and capstone design. While the capstone course was already receiving information literacy instruction, the professional ethics course historically had not received instruction and is a course targeted for instruction. Due to the nature of the material in the class, this course is also a new target for introducing technical standards as they apply to case studies in the class.

TECHNICAL STANDARDS INSTRUCTIONAL DEVELOPMENT TOOLKIT

When planning technical standards information literacy instruction, here are some questions you might ask yourself, a colleague, or the course instructor:

- Have the students received information literacy instruction before?
- If so, where are students already receiving information literacy instruction? What topics already have been covered?
- What discipline and level will students receive instruction?
- What is the outcome of the instruction?
- Does this course need face-to-face technical standards instruction, online instruction, or learning object?
- If the instruction is face-to-face or online synchronous, how much time is allotted for instruction?
- When would technical standards instruction make the most sense for the students?
- Will the students be able to access the standards needed for the course, either through library resources or online resources?
- What type of assessment will fit the instructional method, informal or formal?
- What information do you want to gather from the assessment? Teaching techniques or student learning?

TABLE 7.1. *Electrical Engineering Curriculum Mapping Example*

Electrical Engineering

Course Name	English 101	History 101	Engr 102	EE 302: Professional Ethics	Capstone
Instructional Method	Classroom, Assignment, LibGuide	Tutorial, Assignment, LibGuide	LibGuide	Target for Instruction	Classroom
Learning Outcomes					
Understand the concepts of search terms and search strategies	Introduce	Introduce			Reinforce
Apply search terms and techniques to locate and select appropriate sources	Introduce	Introduce	Enhance		Reinforce
Identify critical databases and other resources to create effective search strategies	Introduce	Introduce			Enhance
Understand discipline-specific complexities of authority, currency, credibility, and literature gaps			Introduce		Enhance
Understand, identify, and locate technical standards				Introduce	Enhance

TECHNICAL STANDARD INFORMATION LITERACY ASSESSMENT

When planning instruction, the last step is to close the loop by moving from planning and implementing technical standards information literacy instruction to assessing the learning and teaching to inform future instruction. Many instructors, faculty, and librarians focus on the planning and implementation stages of education without the assessment component due to a few barriers to adding assessment, which might include time, deciding the type of assessment, and what to do with the data once it has been collected. Through assessment, student learning can be measured to see if the course meets the goals and outcomes of the class, program, or institution [28]. Assessment data from technical standards information literacy instruction can be shared with others regarding the skills students learn and can better inform future instructional methods.

Traditional assessment of student learning is usually done through grading assignments over an academic semester or quarter; however, with information literacy assessment, assessment methods can measure student learning beyond the traditional grading system [23]. When planning for assessment, either within a whole course or a class session, it is essential to look at the more considerable outcomes of the program or the university. Some universities have information literacy within the goals for undergraduate students, or it might be a part of the program accreditation; for more information, see Chapters 9–13, which cover standards in specific disciplines.

When starting to design an assessment method, the first step is determining the outcomes to be measured informally or formally, as mentioned above in the backward instructional design. Informal assessment can be observations or reflections. Formal assessments can include surveys, quizzes, or performance reviews. While assessment can be at the course level, programmatic level, or institutional level, this section focuses on classroom-level assessment as it relates to a specific class period or several class periods. See Table 7.2 for the different types of assessment and the data that is collected through each method.

For examples of technical standards information literacy assessments, see the case studies in Part IV.

TABLE 7.2. *Types of Information Literacy Assessment*

Type of Assessment	Description	Data Collected
Informal Observations	Observation of students as they receive instruction or work on an in-class activity. Informal observations can be done spontaneously without any planning.	Real-time feedback regarding student understanding of topics presented during instruction. Instructors can note what went well or consider changes for future instruction sessions.
Informal Questions	Ask students about instruction or questions they have at the end.	Immediate feedback from students. Can identify areas students struggle with understanding.
Minute Paper	Reflection activity for students to reflect on the instruction. Might include a prompt from the instructor such as: What is one thing you learned today? What is one question you still have after today? What do you wish the instructor would have spent more time presenting?	Minute papers can be done anonymously or with identifying information, giving immediate feedback to the instructor regarding the session and allowing for possible follow-up with students regarding questions.
Surveys	Use after instruction to gather information from students regarding the technical standards information literacy instruction.	Used one time or overtime to collect longitudinal data.
Quizzes	Quizzes can be used in multiple types of instruction, including face-to-face, online, and online learning tools.	Used to collect data after instruction and as a pretest before instruction to measure student understanding of concepts.
Performance Review	Assessment of a student's application of skills based on a student product such as a research paper, presentation, poster, etc.	Measures students' understanding and application of skills.

ADVOCATING FOR TECHNICAL STANDARDS INSTRUCTION

As mentioned in the introduction of this chapter, the need for technical standards education is important for students' lifelong learning and applications within professions. When planning for technical standards education, there are different groups that one might need to advocate to such as faculty, administration, or librarians. Below are tips when advocating for technical standards information literacy instruction:

- Start with understanding the technical standards tools and resources that are available at your institution
- Familiarize yourself with the curriculum for departments that might be a candidate for instruction
- Familiarize yourself with the accreditation standards for the discipline (see Chapters 9–13 for specifics)
- Reach out to faculty who are teaching design- or project-based courses at the beginning of the semester
- Cite the research regarding the use of standards outside of academia and the importance of technical standards education in academia
- Offer different formats of instruction to fit the needs of the course and time allocated

COMMON CHALLENGES WHEN ADVOCATING FOR TECHNICAL STANDARDS INFORMATION LITERACY

There isn't time for technical standards education within our already filled curriculum. When this is the question posed by a colleague or instructor, this would be an opportunity to start a conversation about the course materials, topics covered, and potential collaborations. Talking points also can include different types of support or instruction provided to the students without taking up course time.

My students already know how to use the library and find standards. While many students receive information literacy and library instruction throughout their academic careers, some topics to discuss would be the different topics that students are introduced to in other courses through curriculum mapping of information literacy topics in a specific discipline. Curriculum mapping can show the progression of information literacy throughout their academic careers. Instruction at this time can help reinforce the students' information literacy skills and is an opportunity to explore the tools and resources they need for the specific course more deeply.

I have tried to find standards through the library, but they never have what I need. When access to standards is a concern, it can be a great time to discuss how the library collects technical standards or to reach out to a librarian to discuss access to technical standards. For more information regarding access to technical standards, see Chapter 5, and for more information on technical standards and collection development, refer to Chapter 6.

REFERENCES

1   T. Cooklev, "The role of standards in engineering education," in *Innovations in Organizational IT Specification and Standards Development*, IGI Global, 2013, pp. 129–137.

2   B. Harding and P. McPherson, "Incorporating standards into engineering and engineering technology curricula: it's a matter of public policy," presented at the Proceedings of the American Society for Engineering Education Annual Conference and Exposition, June 2009. https://peer.asee.org/5204

3   K. Krechmer, "Teaching standards to engineers," *International Journal of IT Standards and Standardization Research*, 5(2), 2007.

4   B. S. Kunst and J. R. Goldberg, "Standards education in senior design courses," *IEEE Engineering in Medicine and Biology Magazine*, 22(4), 2003, pp. 114–117. https://doi.org/10.1109/MEMB.2003.1237511.

5   J. P. Olshefsky, "The role of standards education in engineering curricula," presented at the ASEE Conference, Mid-Atlantic Section, 2008. http://www.astm.org/studentmember/PDFS/Role_of_Standards.pdf

6   A. S. Khan, A. Karim, and J. A. McClain, "The state of the use of standards in engineering and technology education," presented at the Proceedings of the American Society for Engineering Education Annual Conference and Exposition, June 2013. https://peer.asee.org/22618.pdf

7   J. Gbur and D. Solomon, "Promoting technical standards education in engineering," presented at the 2016 ASEE Annual Conference, June 2016. https://doi.org/10.18260/p.26005.

8   S. I. Rooney and J. S. Stephens-Epps, "Incorporating engineering standards through-out the biomedical engineering curriculum," presented at the ASEE Annual Conference and Exposition, Conference Proceedings, 2019.

9   S. Howe and J. Goldberg, "Engineering capstone design education: Current practices, emerging trends, and successful strategies," in *Design Education Today: Technical Contexts, Programs and Best Practices*, D. Schaefer, G. Coates, and C. Eckert, eds. Cham: Springer International Publishing, 2019, pp. 115–148. https://doi.org/10.1007/978-3-030-17134-6_6.

10 M. Phillips, H. Howard, A. Vaaler, and D. E. Hubbard, "Mapping industry standards and integration opportunities in business management curricula" (2019). Libraries Faculty and Staff Scholarship and Research. Paper 220. https://doi.org/10.1080/08963568.2019.1638662

11 A. Lampousis, "On the pursuit of relevance in standards-based curriculum development: The CCNY approach," *Standards Engineering*, 69(4), 2017, pp. 1, 3–6.

12 M. Phillips and P. McPherson, "Using everyday objects to engage students in standards education," 2016, pp. 1–5.

13 W. E. Kelly, "Standards in civil engineering design education," *Journal of Professional Issues in Engineering Education and Practice*, 134(1), 2008. https://doi.org/10.1061/(ASCE)1052-3928(2008)134:1(59).

14 A. L. Lerner, B. H. Kenknight, A. Rosenthal, and P. G. Yock, "Design in BME: Challenges, issues, and opportunities," *Annals Biomedical Engineering, 34*(2), Feb. 2006, pp. 200–208. https://doi.org/10.1007/s10439-005-9032-1.

15 V. Charter, B. L. Hoskins, and S. B. Montgomery, "Understanding the significance of integrating codes and standards into the learning environment," presented at the Proceedings of the American Society for Engineering Education Annual Conference and Exposition, Salt Lake City, UT, June 2018. https://peer.asee.org/31181.pdfe-of-integrating-codes-and-standards-into-the-learning-environment.pdf

16 M. Phillips, M. Fosmire, and P. B. McPherson, "Standards are everywhere: A freely available introductory online educational program on standardization for product development," *Standards Engineering*, May/June 2018.

17 M. B. Spring, "Standards education at the University of Pittsburgh," *Standards Engineering*, November/December 2014, pp. 12–15.

18 S. T.-h. Ku, "Standardization in action: A critical path for translational STEM education at Drexel University," *Standards Engineering*, November/December 2018, pp. 11–17.

19 C. Leachman and C. Pezeshki, "What's standard? Industry application versus university education of engineering standards," presented at the Proceedings of the American Society for Engineering Education Annual Conference and Exposition, June 2015. https://peer.asee.org/25068

20 D. Solomon, Y.-T. T. Liao, and T. Chapin, "Maximizing the effectiveness of one-time standards instruction sessions with formative assessment," Tampa, FL, 2019. https://peer.asee.org/33087

21 J. Jeffryes and M. Lafferty, "Gauging workplace readiness: assessing the information needs of engineering co-op students," *Issues in Science and Technology Librarianship, 69*, Spring 2012. https://doi.org/10.5062/F4X34VDR.

22 C. Klotzbach-Russell, E. M. Rowley, and R. Starry, "Librarians in the LaunchPad: Building partnerships for entrepreneurial information literacy," *Journal of Business and Finance Librarianship, 27*(1), 2022, pp. 41–56. https://doi.org/10.1080/08963568.2021.1982567.

23 D.H. Ziegenfuss and S. LeMire. "Backward design a must-have library instructional design strategy for your pedagogical and teaching toolbox," *Reference and User Services Quarterly, 59*(2), 2019, pp. 107–112.

24 Buchanan, Heidi, Katy Kavanagh Webb, Amy Harris Houk, and Catherine Tingelstad. "Curriculum mapping in academic libraries," *The New Review of Academic Librarianship, 21*(1), 2015, 94–111. https://doi.org/10.1080/13614533.2014.1001413.

25 H. H. Jacobs. *Getting Results with Curriculum Mapping.* Alexandria, VA: Association for Supervision and Curriculum Development, 2004.

26 F. Salisbury and L. Sheridan. "Mapping the journey: Developing an information literacy strategy as part of curriculum reform." *Journal of Librarianship and Information Science, 43*(3), 2011, pp. 185–93. https://doi.org/10.1177/0961000611411961.

27 N. Cuevas, A. Matveev, and M. Feit. "Curriculum mapping: An approach to study coherence of program curricula," *Department Chair, 20*(1), 2009.

28 C. J. Radcliff, *A practical guide to information literacy assessment for academic librarians.* Westport, CT: Libraries Unlimited, 2007.

FURTHER READING

A. Brown, *The essentials of instructional design: connecting fundamental principles with process and practice*, 4th edition. New York, NY: Routledge, 2020.

C. J. Radcliff, *A practical guide to information literacy assessment for academic librarians.* Westport, CT: Libraries Unlimited, 2007.

D. Warner, *A disciplinary blueprint for the assessment of information literacy.* Westport, CT: Libraries Unlimited, 2008.

G. P. Wiggins and J. McTighe, *Understanding by design.* Alexandria, VA: Association for Supervision and Curriculum Development, 1998.

T. Y. Neely and B. R. Hannelore, *Information literacy assessment: Standards-based tools and assignments.* Chicago: American Library Association, 2006.

8

Standards Educational Resources

Chelsea Leachman, Washington State University

W hen using technical standards in an educational setting, there are many different sources of information that can assist students and educators. This section will discuss standards educational resources, including those from standards developing organizations, academia, the U.S. government, and other international resources. These resources include tutorials, videos, educational programs, and certificate programs. Depending on the program or supplemental resources, these educational resources can be a complete curriculum.

STANDARDS DEVELOPING ORGANIZATIONS

Standards developing organizations (SDOs) aim to develop, coordinate, and revise technical standards to address the needs of a group of affected adopters. In addition to creating standards, many SDOs also provide educational resources regarding their individual histories, development, or use of their standards. The resources listed in this area will be from the lens of the specific standards developing organization, and some are located behind paywalls available only to members. Below are resources from different organizations with a short explanation of the resource:

ANSI Education and Training in Standardization
https://www.ansi.org/education/standards-education-training
The American National Standards Institute (ANSI) provides educational resources for all students, from primary students through professionals. At the lower levels, ANSI offers educational presentations and handouts for educators to use in the classroom. At higher levels, students and professionals can participate in online courses and webinars on standards basics, understanding their specific designation as an American National Standard, the general development process, and the specific ANSI development process. Courses and webinars can both be offered for free or at a cost to the user. The ANSI resources are a good start for professionals new to using technical standards.

American Society for Mechanical Engineers (ASME) Student Resources
https://www.asme.org/codes-standards/training-and-events/engineering-student-resources
The American Society of Mechanical Engineers (ASME) provides resources for students, faculty, instructors, and librarians to learn or teach standards. The resources provided include articles about standards history and development, and case studies.

ASTM International Classroom for Members
https://www.astm.org/products-services/training-courses/member-training.html
ASTM International (formerly known as the American Society for Testing and Materials) develops and publishes voluntary consensus technical standards. ASTM Classroom for Members provides resources for both industry professionals, professors, and students. ASTM offers presentations, handouts, videos, and curriculum related to their standards products for professors. ASTM student members can apply for a free student membership to have access to webinars, competitions, and electronic editions of ASTM news resources. These resources from ASTM can be helpful when creating LibGuides, a content management system used by many libraries, or course space for students in an online environment where self-directed learning is needed.

Outside of specific resources for faculty and students, ASTM offers modules in English and Spanish on an introduction to standards, the development process, and intellectual property. For professionals, ASTM has eLearning courses available for purchase for those wanting to learn about a specific set of standards. To learn more about eLearning courses visit, https://www.astm.org /products-services/training-courses.html.

Institute of Electrical and Electronics Engineers (IEEE) Standards University
https://www.standardsuniversity.org
Through the Institute of Electrical and Electronics Engineers (IEEE) Standards University, faculty and students have access to a wide variety of free resources, including videos, recorded presentations, news articles, and workshops. Videos range in length from a few minutes to longer recorded presentations.

For a fee through IEEE, an innovative approach to standard education is through a standard game developed by IEEE called Mars Space Colony: A Game of Standardization. The game was designed to include standards development and case studies on standards. Participants work in groups reflecting different stakeholders through the standardization process.

National Institute of Standards and Technology (NIST) Standards Services Curricula Development Cooperative Agreement Program
https://www.nist.gov/standardsgov/nist-standards-coordination-office-curricula -development-cooperative-agreement-o
Since 2012, the National Institute of Standards and Technology (NIST) has awarded university programs with funds to help integrate standards and standardization content into the academic curriculum at all levels through curriculum development. Each project ranges from $25,000 to $75,000 for up to 24 months. This site provides the most current news regarding this program, including funding opportunities and awards. Educational resources created by individual institutions can be accessed through links provided on the NIST website, or by contacting the award recipients.

Society for Standards Professionals (SES)
https://www.ses-standards.org/page/StandardsEducationCourses
The audience for the Society for Standards Professionals (SES) resources is aimed at professionals, students, faculty, and consumers. Resources are available for SES members through their website. Some topics that have been covered in the past include fundamentals of standards and conformity assessment, a guide to standards, and courses targeted at specific industries.

ACADEMIC

Academic institutions have created independent technical standards educational resources that are not tied to specific standards developing organizations. Librarians, faculty, or collaborators create many academic institutions' resources. While independent of SDOs, these resources are tied to a specific educational institution. Below are resources from different academic institutions with a short description.

Intellectual Property Shield
https://www.ip-shield.com/nist.aspx
Created by SES and Purdue University, IP Shield provides four short video case studies relating standards to on-the-job scenarios and a self-paced, online course (60 minutes) relating standards to real-world uses in science, technology, engineering, and mathematics (STEM) areas. They also offer a fee-based online standards education course, Standards Aware.

Purdue University Standards are Everywhere Tutorials
https://guides.lib.purdue.edu/NIST_standards
Consisting of five tutorials, the "Standards are Everywhere" project includes an introduction to standards, anatomy of a standard, discovering and locating standards, and an explanation of how standards are related to everyday objects.

GOVERNMENT

Government agencies provide training opportunities and resources for professionals, academia, and the public regarding the standards activities in the government, background materials on standards, and information for the general public.

National Institute of Standards and Technology (NIST)
https://www.nist.gov/standardsgov
Established in 1901, the National Institute of Standards and Technology (NIST) is part of the U.S. Department of Commerce and is one of the nation's oldest physical science laboratories. NIST is responsible for supporting fields such as energy, cybersecurity, climate, and public safety, to name a few. In addition, the Standards Coordination Office (SCO) at NIST oversees www.standards.gov, which provides the following types of resources regarding standards education:

- Training for government agencies
- Curricula development grant for the program
- Summer undergraduate research fellowship opportunities
- Support through the Standards Information Center
- Adventure in Standards board game

INTERNATIONAL

ISO Consumer and Standards: Partnership for a Better World
https://www.iso.org/sites/ConsumersStandards/index.html
ISO created a freely available read-through module for professionals to everyday citizens interested in understanding standards basics, standards development, the ISO standards system, the benefits of standards, and consumer participation in the standards process. At the end of each section, there are self-assessment review questions. The module takes about a half-day to complete and includes many additional resources for users to return to later.

SUMMARY

Above are select educational resources to assist in either personal education or use within the classroom. The resources range from basic information for new standards users to continuing education for professionals. New resources will continue to become available from standards developing organizations, academic institutions, or individuals through feedback from users and the identification of developing needs.

9

Standards in Engineering and Engineering Technology

Margaret Phillips, Purdue University

INTRODUCTION

Engineering and engineering technology educators teach students about standards to meet accreditation requirements and to prepare students for their next steps after graduation. In industry, new hires utilize standards on the job and may even participate in the creation of standards by representing their companies on standards developing committees. In this section, I examine accreditation options for engineering and engineering technology programs in the United States and explore how standards appear in the accreditation language, both directly and indirectly. Additionally, I introduce employer expectations of new hires with regard to standards knowledge.

ACCREDITATION AND CURRICULUM REQUIREMENTS

ABET Accreditation

ABET, formally known as the Accreditation Board for Engineering and Technology, is a major accrediting body of engineering, engineering technology, computing, and applied and natural science programs [1]. Currently, ABET accredits over 4,300 programs in 41 countries [2]. Many engineering and engineering technology students pursue degrees from ABET-accredited programs in order to qualify for future jobs [3], to meet graduate school admission preferences [4], and to fulfill the eligibility requirements of professional licensure [5]. ABET distinguishes "engineering" and "engineering technology" by a program's curricular focus and the career paths of its graduates [6]. In general, engineering programs and career paths are more theoretical, and engineering technology programs and career paths more applied. ABET accreditation criteria for both engineering and engineering technology programs contain multiple direct and indirect connections to standards.

For engineering programs, ABET's 2020-21 Engineering Accreditation Commission (EAC) criteria *Criterion 5: Curriculum* specifies that "the curriculum must include . . . d) a culminating major engineering design experience that 1) incorporates appropriate engineering standards" [7]. Additionally, Table 9.1 identifies multiple indirect connections between selected EAC *Criterion 3: Student Outcomes* and standards. Lastly, the ABET Program Criteria for "Architectural and Similarly Named Engineering Programs" directly specify that the curriculum "includes computer-based technology and considers applicable codes and standards" [7].

For engineering technology programs, ABET's 2020-21 Engineering Technology Accreditation Commission (ETAC) criteria *Criterion 5: Curriculum* (under "Discipline Specific Content") states that the curriculum must "include design considerations appropriate to the discipline and degree level such as: industry and engineering standards and codes" [11]. Additionally, like engineering, ETAC *Criterion 3: Student Outcomes* contains several indirect connections to standards (see Table 9.2).

Additionally, ABET ETAC includes language directly related to standards and/or codes in their program level criteria for 13 programs. See Table 9.3 for details.

TABLE 9.1. *Indirect Connections Between Selected ABET EAC Criterion 3: Student Outcomes and Standards*

Criterion 3: ABET EAC Student Outcomes	Indirect Connection to Standards
(2) "an ability to apply engineering design to produce solutions that meet specified needs with consideration of public health, safety, and welfare, as well as global, cultural, social, environmental, and economic factors"	ABET EAC's definition of engineering design mentions codes and standards: "Engineering design is a process of devising a system, component, or process to meet desired needs and specifications within constraints. . . . For illustrative purposes only, examples of possible constraints include . . . codes, . . . regulations, . . . standards, sustainability, or usability."
(4) "an ability to recognize ethical and professional responsibilities in engineering situations and make informed judgments, which must consider the impact of engineering solutions in global, economic, environmental, and societal contexts"	Standards streamline processes, reduce costs, ensure safety and quality, and promote interoperability [8], which can help engineering students in making informed judgments.
(6) "an ability to develop and conduct appropriate experimentation, analyze and interpret data, and use engineering judgment to draw conclusions"	Many standards, such as ASTM standards, focus on testing materials, products, procedures, and/or processes and can be used to design, conduct, and evaluate experiments. Additionally, multiple data standards exist to facilitate consistent and accurate data interpretation, use, and sharing [9].
(7) "an ability to acquire and apply new knowledge as needed, using appropriate learning strategies"	Standards are one type of knowledge design teams need to draw upon to develop safe and legal solutions to engineering problems [10].

TABLE 9.2. *Indirect Connections Between Selected ABET ETAC Criterion 3:* Student Outcomes *and Standards*

Criterion 3: ABET ETAC Student Outcomes	Indirect Connection to Standards
(2) "an ability to design solutions for well-defined technical problems and assist with the engineering design of systems, components, or processes appropriate to the discipline" (associate degree programs) (2) "an ability to design systems, components, or processes meeting specified needs for broadly-defined engineering problems appropriate to the discipline" (baccalaureate degree programs)	Standards are one type of knowledge design teams need to draw upon to develop safe and legal solutions to engineering problems [10].
(3) "an ability to identify and use appropriate technical literature" (both associate and baccalaureate degree programs)	Standards are one type of knowledge design teams need to draw upon to develop safe and legal solutions to engineering problems [10].
(4) "an ability to conduct standard tests, measurements, and experiments and to analyze and interpret the results" (associate degree programs) (4) "an ability to conduct standard tests, measurements, and experiments and to analyze and interpret the results to improve processes" (baccalaureate degree programs)	Many standards, such as ASTM standards, focus on testing materials, products, procedures, and/or processes and can be used to design, conduct, and evaluate experiments. Additionally, multiple data standards exist to facilitate consistent and accurate data interpretation, use, and sharing [9].

TABLE 9.3. *Direct Connections Between ABET ETAC Program Criteria and Standards and Codes*

ABET ETAC Program	Program Criteria - Connection to Standards
Aeronautical Engineering Technology and Similarly Named Programs	"assembly and support processes, industry standards, regulations and documentation, and computer-aided engineering graphics with added technical depth in at least one of these areas" (associate degree programs)
Chemical/Refinery Process Engineering Technology and Similarly Named Programs	"operating principles (including testing and troubleshooting) of chemical processes and equipment in accordance with applicable safety (including process hazards), health and environmental standards" (both associate and baccalaureate degree programs)
Computer Engineering Technology and Similarly Named Programs	"application of electric circuits . . . local area networks, and engineering standards to the building, testing, operation, and maintenance of computer systems and associated software systems" (both associate and baccalaureate degree programs)
Construction Engineering Technology and Similarly Named Programs	"utilization of techniques that are appropriate to administer and evaluate construction contracts, documents, and codes" (both associate and baccalaureate degree programs)
Electrical/Electronic(s) Engineering Technology and Similarly Named Programs	"application of circuit analysis and design . . . and engineering standards to the building, testing, operation, and maintenance of electrical/electronic(s) systems" (both associate and baccalaureate degree programs)

TABLE 3. *Continued*

Electromechanical Engineering Technology and Similarly Named Programs	"application of statics, dynamics (or applied mechanics), strength of materials, engineering materials, engineering standards, and manufacturing processes to aid in the characterization, analysis, and troubleshooting of electromechanical systems" (both associate and baccalaureate degree programs)
Engineering Graphics/Design/Drafting Engineering Technology (Mechanical) and Similarly Named Programs	"use of . . . industry codes, specifications, and standards (ASME, ANSI or others)" (both associate and baccalaureate degree programs)
Environmental Engineering Technology and Similarly Named Programs	"roles and responsibilities of public and private organizations pertaining to environmental regulations, including applicable standards" (both associate and baccalaureate degree programs)
Fire Protection Engineering Technology and Similarly Named Programs	"codes and standards for life and fire safety" (baccalaureate degree programs)
Healthcare Engineering Technology and Similarly Named Programs	"information technology principles applied to medical equipment systems, including data security and privacy standards" (associate degree programs) "the clinical application of computer networks, . . . including data security and privacy standards" (baccalaureate degree programs)
Instrumentation and Control Systems Engineering Technology and Similarly Named Programs	"communicating the technical details of control systems using current techniques and graphical standards" (both associate and baccalaureate degree programs)

TABLE 3. *Continued*

Mechanical Engineering Technology and Similarly Named Programs	"basic familiarity and use of industry codes, specifications, and standards" (associate degree programs) "Application of industry codes, specifications and standards" (baccalaureate degree programs)
Telecommunications Engineering Technology and Similarly Named Programs	"application of electric circuits, . . . and engineering standards," (both associate and baccalaureate degree programs)

ATMAE ACCREDITATION

The Association of Technology, Management, and Applied Engineering (AT-MAE) is an organization that currently accredits over 465 associate, baccalaureate, and master's degree programs in areas related to technology, management, and applied engineering in the United States [12]. Unlike ABET, ATMAE does not specify particular student learning outcomes [13]. Rather, ATMAE allows individual programs flexibility in specifying their program- and student-level learning outcomes. Some applied engineering programs, such as the University of Texas at Tyler's Industrial Technology Bachelor of Science program, develop program outcomes directly related to standards [14]. Additionally, other programs, such as San Jose State University's Industrial Technology Bachelor of Science program, list sets of standards (e.g., ASTM, ISO, UL) as lab resources for students in their ATMAE accreditation self-study document [15].

EMPLOYERS EXPECTATIONS

Multiple studies convey industry expectations about employee standards knowledge and use. Harding and McPherson [16] surveyed employers in engineering and technology fields and found the majority of respondents (58%) believe new employees should have fundamental knowledge about standards development, as well as basic skills in finding and applying standards, at the time of hire.

Additionally, Jeffryes and Lafferty [17] surveyed engineering co-op students at the University of Minnesota and reported standards were the type of information students needed to find most frequently during their workplace experiences. Nearly 80% of the co-op student survey respondents indicated they were required to find standards, followed by books (61%) and technical reports (53%). Only 33% of respondents reported a need to find scholarly articles during their co-op experiences.

Likewise, Phillips et al. [18] surveyed professional engineers at Caterpillar, Inc., and found that almost 70% of respondents reported using standards for their work. Additionally, Phillips, Zwicky, and Lu [19] reported that standards and codes were the most frequently mentioned information source type in engineering and engineering technology job ads for entry-level hires. Lastly, in a survey of corporate engineering firm "principals" (defined as owners or senior management), Napp [20] found that engineers use standards (at 92.4%) more frequently than other information types.

SUMMARY

Standards are integrated into engineering and engineering technology curricula for two reasons: (1) to meet direct and indirect accreditation requirements, and (2) to best prepare students for the workplace. In the United States, ABET, the main accrediting body for engineering and engineering technology programs, contains extensive language related to standards in their accreditation criteria. Additionally, ATMAE accredits a large number of engineering technology programs. While ATMAE does not directly require standards integration, many ATMAE-accredited programs demonstrate the achievement of student outcomes by providing examples of standards education in curricula.

Curricular integration of standards helps prepare students to meet employer expectations before (e.g., internships, co-ops) and after graduation, as standards have been widely reported as the most frequently used type of information in industry experiences. Educators seeking instructional ideas can consult the multiple case studies of standards curricular integration we share in Part IV of this book.

REFERENCES

1   ABET, "About ABET." https://www.abet.org/about-abet (accessed Nov. 5, 2020).

2   ABET, "Accreditation." https://www.abet.org/accreditation (accessed Nov. 5, 2020).

3   Bureau of Labor Statistics, U.S. Department of Labor, "Occupational outlook handbook: Mechanical engineers." https://www.bls.gov/ooh/architecture-and -engineering/mechanical-engineers.htm#tab-4 (accessed Nov. 5, 2020).

4   "Admissions - School of Engineering - Master of Science in Engineering Management, UAB." https://www.uab.edu/engineering/msem/academics/admissions (accessed Nov. 5, 2020).

5   Maine State Board of Licensure for Professional Engineers, "College students." https://www.maine.gov/professionalengineers/students/college.html (accessed Nov. 5, 2020).

6   ABET, "What programs does ABET accredit?" https://www.abet.org/accreditation /what-is-accreditation/what-programs-does-abet-accredit (accessed Jan. 27, 2021).

7   ABET, "Criteria for accrediting engineering programs, 2020–2021." https://www .abet.org/accreditation/accreditation-criteria/criteria-for-accrediting-engineering -programs-2020-2021 (accessed Nov. 5, 2020).

8   D. C. Thompson, *A guide to standards*, rev. 3rd ed. Portsmouth, NH: Standards Engineering Society, 2011.

9   "Data Standards." https://www.usgs.gov/products/data-and-tools/data-manage ment/data-standards (accessed Nov. 9, 2020).

10 B. Osif, "Make it safe and legal," in *Integrating Information into the Engineering Design Process*, D. Radcliffe and M. Fosmire, eds. West Lafayette, IN: Purdue University Press, 2014, pp. 115–124.

11 ABET, "Criteria for accrediting engineering technology programs, 2020–2021." https://www.abet.org/accreditation/accreditation-criteria/criteria-for-accrediting -engineering-technology-programs-2020-2021 (accessed Nov. 9, 2020).

12 The Association of Technology, Management, and Applied Engineering (ATMAE), "ATMAE accreditation." https://www.atmae.org/page/accreditation (accessed Dec. 7, 2020).

13 The Association of Technology, Management, and Applied Engineering (ATMAE), "2021 accreditation handbook." https://cdn.ymaws.com/www.atmae.org/resource /resmgr/accred_2018/2021_accreditation_handbook_.pdf (accessed Dec. 7, 2020).

14 Department of Technology, The University of Texas at Tyler, "2019 self-study accreditation report for the bachelor of science in industrial technology at The University of Texas at Tyler." https://www.uttyler.edu/soules-college-of-business/technology /tech/files/2019_ATMAE_Self-study_March_13_2019.pdf (accessed Dec. 7, 2020).

15 Department of Aviation and Technology, San Jose State University, "Self-study report 2017 accreditation." http://www.atmae.org/resource/resmgr/docs/4-Year_Program _Accreditation.pdf (accessed Dec. 7, 2020).

16 B. Harding and P. McPherson, "What do employers want in terms of employee knowledge of technical standards and the process of standardization?," presented at the ASEE Annual Conference and Exposition, 2010. https://doi.org/10.18260/1-2 --16474.

17 J. Jeffryes and M. Lafferty, "Gauging workplace readiness: Assessing the Information needs of engineering co-op students," 69, 2012. https://doi.org/10.5062/F4X34 VDR.

18 M. Phillips, M. Fosmire, L. Turner, K. Petersheim, and J. Lu, "Comparing the Information needs and experiences of undergraduate students and practicing engineers," *Journal of Academic Librarianship, 45(1)*, Jan. 2019, pp. 39–49. https://doi .org/10.1016/j.acalib.2018.12.004.

19 M. Phillips, D. Zwicky and J. Lu, "Initial Study of information literacy content in engineering and technology job postings," 2020 IEEE Frontiers in Education Conference (FIE), Uppsala, Sweden, 2020, pp. 1–3. https://doi.org/10.1109/FIE44824.2020 .9274195.

20 J. B. Napp, "Survey of Library services at engineering news record's top 500 design firms: Ten years later," *Science and Technology Libraries*. https://www.tandfonline .com/doi/full/10.1080/0194262X.2017.1349715 (accessed Dec. 8, 2020).

10

Standards in Computer Science and Information Technology

Daniela Solomon, Case Western Reserve University

INTRODUCTION

The speed of information technology (IT) development over the last 50 years has been unprecedented in the history of humanity and has resulted in IT becoming ubiquitous, impacting every aspect of daily activities worldwide. The explosion of the Internet and wireless networking changed the way people work, communicate, and live. The development was sustained by the extraordinary number of innovations in the field, and was made possible by ensuring interoperability between the myriad information and communication systems on a global scale. In the process, issues such as compatibility, access, privacy, security, and confidentiality were identified as critical to all information and communication technology (ICT) stakeholders. Successful integration of new technologies and applications was enabled by the accelerated evolution of the technical standards in the field. To keep up with the ICT innovations, the standardization process had to find ways to speed up the time it usually takes to finalize a standard and to compromise between achieving general consensus, speed, and quality. It was also necessary to find means to protect the intellectual property rights of innovations while still being able to ensure interoperability. The ICT standards have become so embedded with the existing technology that a high level of compliance is reached even with the voluntary standards. Consequently, knowledge of existing standards and the standards developing organizations (SDOs) creating the various standards applicable to the ICT field is critical for the practicing specialists.

ICT STANDARDS

Technical standards applicable to ICT are numerous and deal with the different aspects of the ICT systems: hardware, software, communication networks, Internet, data, and any of the types of applications developed so far. ICT standards can be established as de facto (PDF, HTML), by public or private organizations through the regular formal process, or as government regulations [1].

The most important functions of ICT standards are [2]:

- Interoperability and compatibility
- Quality
- Variety reduction
- Information and measurement

Hardware standards specify hardware requirements necessary for an intended purpose and assure that the components are interchangeable and compatible with the software. The ever-increasing rate of adoption of new technologies makes interoperability a critical feature for any ICT system. Hardware standards make possible the interconnections between various systems or systems' parts from different vendors without affecting the expected functionality. Software standards assure that various software work with the same accuracy on all available ICT systems. Noncompliance with software standards results in requirements for specific code implementation, which translates into delays, increased costs, and difficulties in updating the systems.

Based on their purpose, ICT standards can be categorized as [3]:

Terminology standards: compile structured vocabularies, terminologies, code sets, and classification systems for ICT systems

Measurements or test methods: define the objectives and guidelines for testing ICT

Specifications: detailed and formal description of a set of characteristics or requirements that are relevant to a specific item

System architecture: support the formal description of ICT systems and their components, characteristics, and processes

Reference models: inform the design of the architecture of ICT systems according to a given model

Software and networking standards: documents about computer software, including programming languages, application programming interfaces (API), communication protocols file information, and formats

Quality assurance: requirements for managing the quality of projects or systems

As the regular standards developing process proved to be too slow for the adoption of ICT technologies and applications, the standardization process had to be changed. While some ICT standards still are developed following the typical process, many ICT-related standards are developed through a modified process that allows reaching the finished document level in a shorter time. Sometimes, however, even this process proved to be too slow, and the documents are still in draft stage when starting over is necessary. This modified process follows open standards principles [4] and consensus is being maximized but not always required [5]. The resulting standards are available to the large public but may have various use rights associated with them. The associated use rights are determined based on whether the standard covers proprietary technology that must be licensed and may result in fees associated with their use [6]. This happens when the proprietary technology is declared a standard essential patent and the owner of the patent agrees to license the patent on fair, reasonable, and nondiscriminatory terms (FRAND); however, if the patent owner is not participating in the standard-setting process, the owner is not obligated to license on FRAND terms [7].

Different organizations have various definitions for open standards and the requirements for the associated fees. For example, the International Telecommunications Union (ITU)'s Telecommunications branch, ITU-T, defines open standards as:

standards made available to the general public and are developed (or approved) and maintained via a collaborative and consensus driven process. Open Standards facilitate interoperability and data exchange among different products or services and are intended for widespread adoption [8].

However, ITU-T standards may have associated fees since many ITU-T standards are standards-essential patents that define the IT industry:

Intellectual property rights (IPRs) essential to implement the standard to be licensed to all applicants on a worldwide, non-discriminatory basis, either (1) for free and under other reasonable terms and conditions or (2) on reasonable terms and conditions (which may include monetary compensation). Negotiations are left to the parties concerned and are performed outside the SDO [8].

On the other hand, the World Wide Web Consortium (W3C), a membership-based international community of organizations, allows royalty-free implementation of their open web standards. Their definition of open standards includes a set of requirements that needs to be met for a standard to qualify as open [9]:

- transparency (due process is public, and all technical discussions, meeting minutes, are archived and referenceable in decision making)
- relevance (new standardization is started upon due analysis of the market needs, including requirements phase, e.g., accessibility, multi-linguism)
- openness (anybody can participate, and everybody does: industry, individual, public, government bodies, academia, on a worldwide scale)
- impartiality and consensus (guaranteed fairness by the process and the neutral hosting of the W3C organization, with equal weight for each participant)
- availability (free access to the standard text, both during development and at final stage, translations, and clear IPR rules for implementation, allowing open source development in the case of Internet/Web technologies)
- maintenance (ongoing process for testing, errata, revision, permanent access)

The debate on whether open standards should include licensing fees or not is ongoing. Other major internationally recognized standards bodies such as the Internet Engineering Task Force (IETF), International Organization for Standardization (ISO), and International Electrotechnical Commission (IEC) also allow for monetary compensation for patent licensing fees, regardless of whether they use the term "open standard" or not.

ACCREDITATION AND CURRICULUM REQUIREMENTS

Accreditation of ICT programs in academia is not as common as it is for other academic programs. In the last few years, however, there has been an increased interest in accreditations of ICT programs, especially if they are part of the engineering schools for which accreditation from the Accreditation Board for Engineering and Technology (ABET) is mandatory. The organization responsible for the accreditation of the ICT programs is the Computing Accreditation Commission (CAC), which is one of the four commissions operated by the ABET [10].

The increased interest in accreditation was determined by the requirements of governmental jobs in ICT fields, professional certifications, and professional licensing for graduates from accredited programs.

The Computing Accreditation Commission has two sets of criteria, *general* and *program-specific*. Each program accredited by an ABET commission must satisfy every criterion included in the general criteria for that commission as well as all program criteria implied by the program title [11]. This commission is accrediting only undergraduate programs, and its criteria are applicable to all computer science, cybersecurity, information systems, and information technology or similarly named computing programs. All ABET commissions recommend topics and skills for graduates but leave institutions the freedom to decide on the specifics of program courses [11].

The general criteria include criteria for eight categories: students, program educational objectives, student outcomes, continuous improvement, curriculum, faculty, facilities, and institutional support. For the purpose of this book, Criterion 3, "Student Outcomes," and Criterion 5, "Curriculum," provide the most information on whether standards education is expected.

Among other requirements, Criterion 3, "Student Outcomes," lists that graduates of the program will have an ability to "design, implement, and evaluate a computing-based solution to meet a given set of computing requirements in the context of the program's discipline" and "recognize professional responsibilities and make informed judgments in computing practice based on legal and ethical principles" [11].

Criterion 5, "Curriculum," includes the requirement that "curriculum must combine technical, professional, and general education components to prepare students for a career, further study, and lifelong professional development in the computing discipline associated with the program" [11].

Despite not mentioning technical standards specifically, these three specific requirements alone represent good arguments for standards education being beneficial to students in ICT programs in order to become valuable professionals. Additional support for this argument comes from Criterion 5, "Curriculum," which mention that computing topics must include [11]:

- Techniques, skills, and tools necessary for computing practice.
- Principles and practices for secure computing.
- Local and global impacts of computing solutions on individuals, organizations, and society.

These criteria refer specifically to privacy, integrity, and security, which are among the top topics of interest for the ICT standards.

The specific program criteria for the computer science programs adds to Criterion 5, "Curriculum," the requirement for "a major project that requires integration and application of knowledge and skills acquired in earlier course work." A successful major project for students should simulate real work experiences as much as possible, including opportunities for students to learn about and use standards in their projects [11].

Program criteria for the cybersecurity and similarly named computing programs adds to Criterion 3, "Student Outcomes," to "apply security principles and practices to maintain operations in the presence of risks and threats," and Criterion 5, "Curriculum," adds that coursework must include:

- Application of the crosscutting concepts of confidentiality, integrity, availability, risk, adversarial thinking, and systems thinking
- Fundamental topics from each of the following [11]:
  1. *Data Security*: protection of data at rest, during processing, and in transit.
  2. *Software Security*: development and use of software that reliably preserves the security properties of the protected information and systems.

3. *Component Security:* the security aspects of the design, procurement, testing, analysis, and maintenance of components integrated into larger systems.

4. *Connection Security:* security of the connections between components, both physical and logical.

5. *System Security:* security aspects of systems that use software and are composed of components and connections.

6. *Human Security:* the study of human behavior in the context of data protection, privacy, and threat mitigation.

7. *Organizational Security:* protecting organizations from cybersecurity threats and managing risk to support successful accomplishment of the organizations' missions.

8. *Societal Security:* aspects of cybersecurity that broadly impact society as a whole.

Protection of national information infrastructure poses many difficult issues and is critical at the national level. Recent cyberattacks on private business and the public have brought to light the need for capable professionals to protect and prevent such problems. Consequently, knowledge of the NIST framework for reducing cyber risks to critical infrastructure [12] and any other existing standards, guidelines, and practices that address cybersecurity-related risks is fundamental to students in these programs and are applicable to all organizations, regardless of their size, industry, or sector. Additionally, working knowledge of the international ISO/IEC 27000 standards series that provide specifications for best-practice information security management is a basic skill that cybersecurity programs graduates should develop [13].

Program criteria for the information systems and similarly named computing programs adds to Criterion 3, "Student Outcomes," the requirement that graduates also will have an ability to "support the delivery, use, and management of information systems within an information systems environment [IS]" [11], and to Criterion 5, "Curriculum," which adds requirements for coursework to include "coverage of fundamentals and applied practice in application development; data and information management; information technology infrastructure; systems analysis, design, and acquisition; project management; and the role of information systems in organizations" [11].

Information systems have become the backbone of all organizations because they support operations, management, and decision making in business processes. In consequence, ensuring their maximum functionality, confidentiality, and integrity is crucial for organizations. Actions taken to facilitate the functionality include, but are not limited to, authentication, data encryption, password security, backups, and firewalls. For the graduates of these programs to be successful in their future careers, it is important then to understand and utilize the widely recognized and adopted international ISO standards for management systems that provide requirements or guidance to help organizations improve their performance [14].

Program criteria for the information technology and similarly named computing program adds to Criterion 3, "Student Outcomes," the ability to "identify and analyze user needs and to take them into account in the selection, creation, integration, evaluation, and administration of computing-based systems [IT]" [11], and to Criterion 5, "Curriculum," which adds the requirement to include "coverage of fundamentals and applied practice" in the following: (a) the core information technologies of human-computer interaction, information management, programming, web systems and technologies, and networking; (b) system administration and system maintenance; and (c) system integration and system architecture [11].

ICT STANDARDS DEVELOPING ORGANIZATIONS

ICT standards are developed at international and national levels with many participating organizations. Most official computer standards are set by one of the following organizations:

International

International Organization for Standardization (ISO)
https://www.iso.org
ISO is the world's largest developer and publisher of international standards, comprising a network of national standards institutes of 157 countries.

International Electrotechnical Commission (IEC)
https://www.iec.ch
IEC is the world's leading organization for the preparation and publication of international standards for all electrical, electronic, and related technologies.

International Telecommunications Union (ITU)
http://www.itu.int
ITU is the leading United Nations agency for information and communication technologies. ITU defines international standards, particularly communications protocols.

Regional

European Standardization Organizations
https://www.cencenelec.eu
CEN, the European Committee for Standardization is one of three organizations responsible for developing and defining standards at the European level. CEN brings together the National Standardization Bodies of the EU countries and supports standardization activities in relation to a wide range of fields and sectors, including air and space, chemicals, construction, consumer products, defense and security, energy, the environment, food and feed, health and safety, health care, ICT, machinery, materials, pressure equipment, services, smart living, transport, and packaging.

CENELEC, the European Committee for Electrotechnical Standardization, another organization at the European level, brings together the National Electrotechnical Committees of EU countries and supports activities in relation to a wide range of fields and sectors, including electromagnetic compatibility, accumulators, primary cells and primary batteries, insulated wire and cable, electrical equipment and apparatus, electronic, electromechanical and electrotechnical supplies, electric motors and transformers, lighting equipment and electric lamps, low voltage electrical installations material, electric vehicles, railways, smart grid, smart metering, solar (photovoltaic) electricity systems, and so forth.

ETSI, the European Telecommunications Standards Institute (https://www.etsi.org), the third organization part of the European Standardization Organizations, supports the timely development, ratification, and testing of globally applicable standards for ICT-enabled systems, applications, and services.

PASC, the Pacific Area Standards Congress
https://pascnet.org
PASC is a voluntary, independent organization of national standards bodies representing countries and territories of the Pacific Rim. PASC is a forum to exchange information and views about international standardization activities and

strengthen positions at the International Organization for Standardization (ISO) and the International Electrotechnical Commission (IEC).

ARSO, the African Organisation for Standardisation
https://www.arso-oran.org
ARSO is Africa's intergovernmental standards body mandated to develop tools for standards development, standards harmonization, and implementation of these systems to enhance Africa's internal trading capacity, increase Africa's product and service competitiveness globally, and uplift of the welfare of African consumers as well as provide a standardization forum for future prospects in international trade referencing.

SARSO, the South Asian Regional Standards Organization
https://sarso.org
SARSO has been established to achieve and enhance coordination and cooperation among SAARC (South Asian Association for Regional Cooperation) member countries in the fields of standardization and conformity assessment, and is aimed to develop harmonized standards for the region to facilitate intraregional trade and to have access to the global market.

COPANT, the Pan American Standards Commission
https://www.copant.org
COPANT is a civil nonprofit association that is the reference for technical standardization and conformity assessment for the countries of the Americas.

Country-Specific: United States

National Institute of Standards and Technology (NIST)
https://www.nist.gov
NIST is a nonregulatory agency that is part of the United States Department of Commerce, the mission of which is to promote American innovation and industrial competitiveness. NIST structure includes several laboratories, including the Communications Technology Laboratory (CTL) and the Information Technology Laboratory (ITL).

American National Standards Institute (ANSI)
http://www.ansi.org
ANSI is a private nonprofit organization that oversees the development of voluntary consensus standards. ANSI creates standards for a wide range of technical areas, including ICT.

STANDARDS INITIATIVES

3rd Generation Partnership Project (3GPP)
www.3gpp.org
3GPP brings together, in a partnership project, SDOs operating in the telecommunication field in countries and regions across the globe. 3GPP covers cellular telecommunications network technologies, including radio access, the core transport network, service capabilities and hooks for nonradio access to the core network, and for interworking with Wi-Fi networks.

Organization for the Advancement of Structured Information Standards (OASIS)
https://www.oasis-open.org
OASIS is "a global nonprofit consortium that works on the development, convergence, and adoption of open standards for cybersecurity, blockchain, Internet of Things (IoT), emergency management, cloud computing, legal data exchange, energy, content technologies, and other areas" [15].

PROFESSIONAL ORGANIZATIONS AND INDUSTRIAL COMMUNITIES

Institute of Electrical and Electronics Engineers Standards Association (IEEE SA)
https://standards.ieee.org
IEEE Standards Association (IEEE SA) is a leading consensus-building organization that nurtures, develops, and advances global technologies, through IEEE. IEEE is the largest nonprofit technical professional organization and a leading developer of standards for a broad range of related technologies.

Internet Engineering Task Force (IETF)

https://www.ietf.org

IETF is "a large open international community of network designers, operators, vendors, and researchers concerned with the evolution of the Internet architecture and the smooth operation of the Internet."

World Wide Web Consortium (W3C)

https://www.w3.org

The W3C is an international community and the main standards organization for the World Wide Web.

SUMMARY

The extraordinary development of the information and communication technologies in the last 50 years brought humanity to the information era. Ensuring interoperability, integrity, privacy, and security of the ICT systems is critical for daily activities. ICT-specific technical standards provide a working framework for these conditions to be met. ICT technical standards are so intertwined with the ICT topics that standards education within the ICT academic programs should be an integral part of the curriculum, despite not being required by the accreditation body.

REFERENCES

1   N. Abdelkafi, R. Bolla, C. J. Lanting, A. Rodriguez-Ascaso, M. Thuns, and M. Wetterwald, *Understanding ICT standardization: Principles and practices*, ETSI, 2018. https://www.etsi.org/images/files/Education/Understanding_ICT_Standardization_Lo ResPrint_20190125.pdf (accessed Jul. 20, 2021).

2   D. Šimunić and I. Pavić, *Standards and innovations in information technology and communications*. Cham, Switzerland: Springer, 2020.

3   H. J. D. Vries, "IT standards typology," in *Advanced topics in information technology standards and standardization research*, vol. 1. Hershey, PA: Idea Group Publishing, 2006.

4   OpenStand Initiative, "The modern standards paradigm—five key principles," *OpenStand*, 2012. https://open-stand.org/about-us/principles (accessed Jul. 20, 2021).

5 R. Rada, "Consensus versus speed," *in Information technology standards and standardization: A global perspective*, K. Jacobs, ed. Hershey, PA: Idea Group Publishing, 2000, pp. 19–34.

6 T. S. Simcoe, "Open standards and intellectual property rights," in *Open innovation: Researching a new paradigm*, H. Chesbrough, W. Vanhaverbeke, and J. West, eds. Oxford: Oxford University Press, 2006, pp 161–183.

7 J. L. Contreras, "Essentiality and standards-essential patents," in *Cambridge handbook of technical standardization law—antitrust, competition and patent law*, J. L. Contreras, ed. Cambridge, UK: Cambridge University Press, 2017, pp. 209–230. https://doi.org/10.1017/9781316416723.016

8 ITU-T, "Definition of 'open standards,'" *ITU*, 2005. https://www.itu.int:443/en/ITU-T/ipr/Pages/open.aspx (accessed Jul. 18, 2021).

9 W3C, "World Wide Web Consortium (W3C)," 2021. https://www.w3.org (accessed Jul. 18, 2021).

10 ABET, "Accreditation commissions," 2021. https://www.abet.org/accreditation (accessed Jul. 18, 2021).

11 ABET, "Criteria for accrediting computing programs, 2020–2021," 2021. https://www.abet.org/accreditation/accreditation-criteria/criteria-for-accrediting-computing-programs-2020-2021 (accessed Jul. 17, 2021).

12 NIST, "Cybersecurity framework," Nov. 12, 2013. https://www.nist.gov/cyberframework (accessed Jul. 17, 2021).

13 ISO, "ISO/IEC 27001—Information security management." https://www.iso.org/isoiec-27001-information-security.html (accessed Jul. 20, 2021).

14 ISO, "Management system standards," https://www.iso.org/management-system-standards.html (accessed Jul. 17, 2021).

15 OASIS, "OASIS—about us," *OASIS Open.* https://www.oasis-open.org/org (accessed Jul. 20, 2021).

11

Standards in Business

Heather Howard, Purdue University

INTRODUCTION

Industry standards are essential for business decisions and have a significant impact as a way to eliminate waste, reduce costs, market products (e.g., for safety, interoperability, and/or environmental), and lessen corporate liability [1]. Additionally, standards and technical regulations have an impact on more than 93% of global trade [2]. Despite their undeniably large impact on business and the economy, standards education is not generally integrated into business education [3]. Due to this outsized impact, it is difficult to pin down every type of standard that relates to business, as business is related to nearly every aspect of industry. This chapter will focus on standards I have encountered most frequently in my work with business owners and entrepreneurs.

TYPES OF STANDARDS IN BUSINESS

Though many categories of standards can apply to business, some relate specifically to business practices and processes.

Accounting and Finance

Standards in accounting are one of the oldest codified business standards, dating back to double-entry bookkeeping in the 15th century [4]. Current U.S. accounting standards are developed by the Financial Accounting Foundation's (FAF) standard-setting boards: the Financial Accounting Standards Board (FASB) and the Governmental Accounting Standards Board (GASB). The standards they collectively create are known as the U.S. Generally Accepted Accounting Principles (GAAP). Transferrable, relevant, and comparable information helps create efficient, robust capital markets [4]. In finance, standards exist to help protect investors and their investments. The Financial Industry Regulatory Authority (FINRA) is a nonprofit organization that works with the Securities and Exchange Commission (SEC) to oversee U.S. broker deals. As part of this, they write and enforce finance standards [5]. If a business wants to bank or trade internationally, it is imperative to be aware of international finance standards and regulations. For example, in the EU, the European system of financial supervision includes three organizations that create standards to regulate banking and finance: the European Banking Authority (EBA), the European Securities and Markets Authority (ESMA), and the European Insurance and Occupational Pensions Authority (EIOPA) [6–8].

Health and Safety

In the United States, all employers must comply with the standards developed and published by the Occupational Safety and Health Administration (OSHA). These standards focus on the areas of general industry, construction, and maritime [9]. Though OSHA is thought of most often regarding factories or construction, they have many standards that apply to an office environment, and any company that hires even one employee who is not an owner is required to follow the regulations. In addition to OSHA, several organizations issue additional mandatory and voluntary safety standards, such as the International Organization for Standardization (ISO), the American Society of Safety Professionals (ASSP), the American National Standards Institute (ANSI), and the European Committee for Standardization (CEN) [10–13].

Human Resources

In the last few years, there has been significant development of standards for human resource management. These standards can help businesses create and benchmark processes and procedures in areas such as organizational culture, occupational health and safety, knowledge management, recruitment and hiring, compensation and benefits, employment relations (e.g., training and development, workplace privacy, discrimination, workplace safety), cost-per-hire, turnover metrics, governance, compliance and ethics, diversity and inclusion, workforce productivity, performance management, succession planning, personnel records management, and more. The main work in human resources standards development is done by ISO who, in 2011, formed the Technical Committee 260 (TC 260) as a group of human resource management experts to design ISO standards in this area. There are currently 33 participating countries with an additional 25 participating as observing members [14].

Information and Cybersecurity

Information and cybersecurity standards work to define processes, procedures, and approaches that can be used to help keep a system secure. This can include accepting online payments, storing customer and employee data, fraud prevention, and preventing network security attacks. There are several agencies that publish standards in this area, including ISO/IEC 27001 Information Security Management, the NIST Cybersecurity Framework, the ISACA Control Objectives for Information Technologies (COBIT), the Center for Internet Security Critical Security Controls, and the Security Standards Council Payment Card Industry (PCI) Data Security Standard [15–19]. Even if a business is not handling their information technology (IT) internally, it is important to know what these standards are in order to hire an outside firm and be confident in their security. Adhering to these standards also can help give customers a peace of mind in knowing that a business is taking the necessary steps to keep their data secure.

International Business

Countries and regions often have different standards creation bodies, though there are many instances of international standards used across the globe. If a company wants to sell a product in an international market, they need to determine what mandatory and voluntary standards exist in that locale. These can include interoperability, safety, environmental, and many more types of standards. For example, if an electronics company wanted to make a product to sell in the United Kingdom, it would need to comply with consumer protection legislation. This includes the Plugs and Sockets etc. (Safety) Regulations 1994, which requires all U.K. electronic appliances be supplied with a BS 1363 plug, referencing a standard put out by the British Standards Institution (BSI) [20].

Marketing

Meeting a standard can help a business with its marketing strategy. Many types of standards apply to this area, including environmental, quality, safety, performance, interoperability, information and cybersecurity, and more. For example, a business could choose to comply with a voluntary environmental standard to be able to advertise their product as "green" or "environmentally friendly" in order to attract customers who value these traits. A business may also choose to comply with a performance standard to market a product's capabilities or durability.

Some industries also have specific advertising standards, often mandatory, that delineate practices such as how to advertise products to certain consumer groups (e.g., children). In the United States, advertising is most often regulated by the Federal Trade Commission (FTC), and in the United Kingdom, standards are written by the Committee of Advertising Practice (CAP) and regulated and enforced by the Advertising Standards Authority (ASA) [21–22]. Examples of nonregulatory advertising standards include the Better Business Bureau (BBB) Code of Advertising and the Standards of Practice of the American Association of Advertising Agencies (AAAA), and the International Chamber of Commerce (ICC) Advertising and Marketing Communications Code [23–25]. Additionally, the International Council for Advertising Self-Regulation (ICAS) publishes the Global Factbook of Advertising Self-Regulatory Organizations annually, which includes a list of these organizations worldwide [26].

Personnel Certification

Many industries include jobs that require personnel certification, such as heavy equipment operators, food handlers, asset management professionals, and computer technicians. These certification programs are based on industry specific standards, and the training programs themselves can be accredited by an accrediting body, such as the ANSI National Accreditation Board (ANAB), to show compliance with the ISO/IEC 17024 standard [27]. Following a standard of certification ensures that professionals in these jobs have an expected set of knowledge, skills, and abilities. This can help a business with hiring, training, and liability.

Quality Management

Quality standards can help businesses lower costs through reduced redundancy, fewer errors or recalls, and reducing the time it takes a product to get to market. Additionally, complying with quality standards can help ensure a product produced in one country can be sold in another, and should be a consideration when determining where a product will be marketed [28]. The most widely known quality standards are likely the ISO 9000 family, which are based on seven quality management principles: customer focus, leadership, engagement of people, process approach, improvement, evidence-based decision making, and relationship management [29–30]. In addition to the ISO, large standards organizations, such as the National Institute of Standards and Technology (NIST), as well as smaller industry specific organizations, such as Spectrum Quality Standards and the International Auditing and Assurance Standards Board, also create quality standards.

STANDARDS WITHIN THE CURRICULUM

The Association to Advance Collegiate Schools of Business (AACSB) is the traditional accrediting body for business schools. The newest 2020 standards are less prescriptive regarding specific topics to be covered in a degree program, and rely on individual schools and programs to develop the content of the curriculum themselves. They require that a curriculum "include relevant competencies that prepare graduates for business careers and foster a lifelong learning mindset" and that "curriculum should reflect current and innovative business theories and practices" [31]. Teaching standards in the business curriculum are clearly appropriate in this framework and would help support the curricular content as specified in the AACSB accreditation standards.

The Accreditation Council for Business Schools and Programs (ACBSP) does not list standards in their curriculum standards for associates, baccalaureate, or doctoral degree programs. They do list some broad categories in which standards education would be an important contribution: marketing, accounting, management, human resources, global dimensions of business, and business policies [32].

There are only a few examples of integrating standards into the business curriculum. Two examples come from San José State and Northwestern University, which received awards from the NIST program that supports the development of learning materials that integrate standards into the engineering and business curricula [33]. Today evidence of neither project at San José State or Northwestern University can be found on the respective university websites. Phillips et al. found little evidence of standards integration in undergraduate management curricula in either the Purdue University Undergraduate Management Program or Texas A&M University Bachelor of Business Administration Program, though the authors did find ample opportunity where it would easily fit [3].

Internationally, few additional examples can be found. The Rotterdam School of Management at Erasmus University hosts an endowed chair in standardization funded by the Netherlands Standardization Institute. The school offers both undergraduate- and graduate-level courses on standardization and includes the option for undergraduates to write a thesis on a standardization topic [34]. The University of Geneva originally launched a master's degree in standardization, social regulation, and sustainability in 2011, and recently rebranded the program to a master's degree in sustainable societies and social change. This program is run through a university partnership with ISO and includes courses on standardization [35].

APPLICATION OF STANDARDS IN PRACTICE

Expanding to International Markets

An example of how standards intersect in multiple ways with business is in the consideration of expanding product distribution into an international market. If a company in the United States is considering expanding the market for a personal electronics product to the United Kingdom, they will need to consider many standards-related questions. If the original product was not designed with U.K. sales in mind, it will be necessary to make sure the product meets any required safety, interoperability, and environmental standards. Will the current connections work with U.K. systems? Does the packaging and labeling meet any required product marking or labeling standards? Are there different testing standards that need to be considered? The company should also assess its competition and determine if any additional voluntary standards exist that might give a competitive advantage. Would following additional quality or environment standards help set them apart? While the cost of standards compliance can be high, the benefit may (or may not) present a favorable return on that investment.

BREWERY ENTREPRENEURSHIP

As the owner of Escape Velocity Brewing, a small brewpub that opened in Lafayette, Indiana, in 2020, I learned firsthand how important standards are to an entrepreneur and small business owner. During construction, we had to comply with the Indiana Building Code that references several different standards. For example, the 2020 Indiana Energy Conservation Code requires compliance to ANSI/ASHRAE 90.1 Energy Standard for Building Except Low-Rise Residential Buildings, 2007 Edition, I-P Edition [36]. The state does not link to or list the requirements of this standard, so business owners need to know how to research this standard on their own. While doing this research, I found that this standard was updated multiple times since the 2007 edition, so I had to make sure the standard I referenced was actually the older version, since that is what is referred to in the Indiana code. Once I found the code, I learned that we were required to build a vestibule in our brewery space. The code states, "Building entrances that separate conditioned space from the exterior shall be protected with an enclosed vestibule, with all doors opening into and out of the vestibule equipped with self-closing devices" [37]. Additionally, the 2014 Indiana Building Code references both the International Building Code, 2012 Edition, First Printing and ANSI A117.1 Accessible and Usable Buildings and Facilities, 2009 Edition, First Printing, which informed additional design and construction decisions [36].

As an operating brewery, there are several environmental standards we need to consider, including both Indiana state and federal water quality standards. The Brewers Association (BA) also has published standards on water and wastewater management to provide sustainable best practices for breweries [38]. We also adhere to the BA Draught Beer Quality Manual that includes standards for the equipment, operation, and maintenance of draught beer systems, and the BA Best Practices Guide to Quality Craft Beer—Delivering Optimal Flavor to the

---

Consumer that includes quality standards for craft beer storage and handling practices [39,40]. When brewing our beer, we reference both the Beer Judge Certification Program Style Guidelines and the Brewers Association Beer Style Guidelines as standards documents that explicitly describe the characteristics of hundreds of beer styles [41–42]. OSHA safety standards are also critical, as a craft brewery is a small production facility with many potential ways we or our employees could get hurt. We want to ensure safety and mitigate risks from things like slips and falls, caustic chemicals used in confined spaces, or improper use of personal protective equipment (PPE).

In addition to our brewery operations, Escape Velocity Brewing also has a full kitchen. Indiana law mandates that we have at minimum one certified food handler on staff [43]. The personnel certification must be provided by an ANSI-CFP Accreditation Program organization to ensure compliance with food safety standards. We also are required to comply with the Indiana Retail Food Establishment Sanitation Requirements, which cover everything from employee hygiene to food storage to kitchen and equipment cleaning and sanitation and more [44].

On the business side of our operation, we use GAAP standards for our business accounting. As we grow and hire more employees, we plan to use the growing set of ISO human resource standards to ensure we are following best practices for our human resource management. The Alcohol and Tobacco Tax and Trade Bureau regulates alcohol beverage advertising in the United States, so we must follow their regulations, but we also consider the standards provided in the Beer Institute Advertising/Marketing Code and Buying Guidelines and the Brewers Association Marketing and Advertising Code when creating our marketing and advertising materials [45–47].

SUMMARY

Because of the nature of business, there are few, if any, types of standards that do not need to be considered in business decisions. Business owners and operators need to know when it is mandatory to comply with a standard, and how to find the standard document referenced by the law or regulation. They need to make determinations regarding complying with voluntary standards and understand the benefits that could come from doing so. Complying with a standard can be an excellent way to attract new customers, encourage innovation, save money and resources, and ensure the health, safety, and well-being of employees. Understanding standards also can help a business make informed purchasing decisions. It is important for a business to determine if they should try to become part of the standards creation process by having employees join standards creation bodies or committees. While there are costs to doing this type of work, it may be advantageous to have a voice at the table in the creation of standards that may end up influencing the business.

REFERENCES

1   D. C. Thompson, *A guide to standards*. Portsmouth, NH: Standards Engineering Society, 2011.

2   J. Okun-Kozlowicki, "Standards and regulations: Measuring the link to goods trade," p. 32, Jun. 2016.

3   M. Phillips, H. Howard, A. Vaaler, and D. E. Hubbard, "Mapping industry standards and integration opportunities in business management curricula," *Journal of Business and Finance Librarianship, 24(1–2)*, Apr. 2019, pp. 17–29. https://doi.org/10.1080/08963568.2019.1638662.

4   Financial Accounting Foundation, "Accounting standards." https://www.accountingfoundation.org/jsp/Foundation/Page/FAFSectionPage&cid=1351027541272 (accessed Nov. 22, 2021).

5   The Financial Industry Regulatory Authority, Inc., "FINRA." https://www.finra.org (accessed Dec. 16, 2021).

6   European Banking Authority, "The single rulebook." https://www.eba.europa.eu/regulation-and-policy/single-rulebook (accessed Dec. 16, 2021).

7   European Insurance and Occupational Pensions Authority, "Technical standards." https://www.eiopa.europa.eu/document-library/technical-standards-0_en (accessed Dec. 16, 2021).

8   European Securities and Markets Authority, "Guidelines and technical standards." https://www.esma.europa.eu/convergence/guidelines-and-technical-standards (accessed Dec. 16, 2021).

9   Occupational Safety and Health Administration, "Law and regulations." https://www.osha.gov/laws-regs (accessed Dec. 15, 2021).

10 ANSI, "Workplace safety standards." https://webstore.ansi.org/industry/safety-workplace (accessed Dec. 15, 2021).

11 American Society of Safety Professionals, "Safety standards topics." https://www
    .assp.org/standards/standards-topics (accessed Dec. 15, 2021).

12 ISO, "ISO 45000 family: Occupational health and safety." https://www.iso.org/iso
    -45001-occupational-health-and-safety.html (accessed Dec. 15, 2021).

13 European Committee for Standardization, "Personal protective equipment."
    https://www.cencenelec.eu/areas-of-work/cen-cenelec-topics/personal-protective
    -equipment (accessed Dec. 15, 2021).

14 ISO, "ISO/TC260: Human resource management," 2021. https://committee.iso.org
    /home/tc260 (accessed Dec. 16, 2021).

15 Center for Internet Security, "CIS critical security controls." https://www.cisecurity
    .org/controls/ (accessed Dec. 16, 2021).

16 ISACA, "COBIT: Effective IT governance at your fingertips." https://www.isaca
    .org/resources/cobit (accessed Dec. 16, 2021).

17 ISO, "ISO/IEC 27001: Information security management." https://www.iso.org
    /isoiec-27001-information-security.html (accessed Dec. 16, 2021).

18 NIST, "Cybersecurity framework." https://www.nist.gov/cyberframework (ac-
    cessed Dec. 16, 2021).

19 PCI Security Standards Council, LLC, "Payment Card Industry (PCI) data security
    standard: Requirements and security assessment procedures, version 3.2.1," 2018.
    https://www.pcisecuritystandards.org/documents/PCI_DSS_v3-2-1.pdf (accessed
    Dec. 16, 2021).

20 "The Plugs and Sockets etc. (Safety) Regulations 1994," 1994. https://www
    .legislation.gov.uk/uksi/1994/1768/made (accessed Dec. 15, 2021).

21 Committee of Advertising Practice, "Advertising codes." https://www.asa.org.uk
    /codes-and-rulings/advertising-codes.html (accessed Dec. 16, 2021).

22 Federal Trade Commission, "Advertising and marketing." https://www.ftc.gov
    /tips-advice/business-center/advertising-and-marketing (accessed Dec. 16, 2021).

23 American Association of Advertising Agencies, "Standards of practice of the
    American Association of Advertising Agencies," 1990. https://ams.aaaa.org/eweb
    /upload/inside/standards.pdf (accessed November 22, 2021).

24 Better Business Bureau, "BBB code of advertising." https://www.bbb.org (accessed
    Dec. 16, 2021).

25 International Chamber of Commerce, "ICC advertising and marketing com-
    munication code," 2018. https://iccwbo.org/content/uploads/sites/3/2018/09/icc
    -advertising-and-marketing-communications-code-int.pdf (accessed Dec. 16, 2021).

26 International Council for Ad Self-Regulation, "2020 global factbook of advertis-
    ing self-regulatory organizations," 2020. https://icas.global/wp-content/uploads
    /2020_Global_SRO_Factbook.pdf (accessed November 22, 2021).

27 ANSI National Accreditation Board, "Accreditation program for personnel certification bodies under ISO/IEC 17024." https://anab.ansi.org/credentialing/personnel-certification (accessed Dec. 15, 2021).

28 American Society for Quality, "What are quality standards?" https://asq.org/quality-resources/learn-about-standards (accessed Nov. 22, 2021).

29 ISO, "Quality management principles," 2015. https://www.iso.org/files/live/sites/isoorg/files/store/en/PUB100080.pdf (accessed November 22, 2021).

30 ISO, "ISO 9000 family: Quality management." https://www.iso.org/iso-9001-quality-management.html (accessed Dec. 16, 2021).

31 AACSB, "2020 guiding principles and standards for business accreditation," 2020. https://www.aacsb.edu/-/media/documents/accreditation/2020-aacsb-business-accreditation-standards-july-2021.pdf?rev=80b0db4090ad4d6db60a34e975a73b1b&hash=D210346C64043CC2297E8658F676AF94 (accessed Dec. 16, 2021).

32 ACBSP, "ACBSP unified standards and criteria for demonstrating excellence in business programs," 2021. https://cdn.ymaws.com/acbsp.org/resource/resmgr/docs/accreditation/Unified_Standards_and_Criter.pdf (accessed Dec. 16, 2021).

33 NIST, "NIST standards services curricula development cooperative agreement program: awardees." https://www.nist.gov/standardsgov/nist-standards-services-curricula-development-cooperative-agreement-program-awardees (accessed Dec. 16, 2021).

34 H. J. de Vries, "The Netherlands: A business approach to standards," 2009. https://www.iso.org/files/live/sites/isoorg/files/archive/pdf/en/iso-award2009-netherlands.pdf (accessed Jan. 16, 2019).

35 Université de Genève, "Sustainable societies and social change." https://www.unige.ch/sciences-societe/formations/masters-in-english/socialchange (accessed Dec. 16, 2021).

36 Indiana Department of Homeland Security, "Rules of the Indiana Fire Prevention and Building Safety Commission," Aug. 30, 2021. https://www.in.gov/dhs/fire-and-building-safety/fpbsc-rules (accessed Dec. 13, 2021).

37 ASHRAE, "ANSI/ASHRAE/IESNA Standard 90.1-2007, Energy Standard for Buildings Except Low-Rise Residential Buildings, SI Edition," 2007.

38 Brewers Association, "Wastewater management guidance manual," 2015. https://www.brewersassociation.org/educational-publications/wastewater-management-guidance-manual (accessed Dec. 16, 2021).

39 Brewers Association, "Best practices guide to quality craft beer—delivering optimal flavor to the consumer," 2014. https://www.brewersassociation.org/attachments/0001/3980/EDP_Quality.pdf (Dec. 16, 2021).

40 Brewers Association, ed., *Draught beer quality manual*, 4th edition. Boulder, CO: Brewers Publications, 2019.

41 BJCP, Inc., "Beer judge certification program 2015 style guidelines," 2015. https:// legacy.bjcp.org/docs/2015_Guidelines_Beer.pdf (accessed Dec. 16, 2021).

42 Brewers Association, "Brewers association 2021 beer style guidelines," 2021. https:// cdn.brewersassociation.org/wp-content/uploads/2021/02/22104023/2021_BA _Beer_Style_Guidelines_Final.pdf (accessed Dec. 16, 2021).

43 Indiana State Department of Health, "Certification of food handler requirements: Title 410 IAC 7-22," 2006.

44 Indiana State Department of Health, "Retail food establishment sanitation re-quirements: Title 410 IAC 7-24," 2004.

45 Beer Institute, "Advertising/marketing code and buying guidelines," 2018. https:// www.beerinstitute.org/wp-content/uploads/2018/12/BEER-6735-2018-Beer-Ad -Code-Update-Brochure-for-web.pdf (accessed Dec. 17, 2021).

46 Brewers Association, "Brewers association marketing and advertising code," 2017. https://cdn.brewersassociation.org/wp-content/uploads/2017/04/BA_Advertising _Code_Overview.pdf (accessed Dec. 17, 2021).

47 TTB, "Advertising—alcohol beverage advertising," 2019. https://www.ttb.gov /advertising/alcohol-beverage-advertising (accessed Dec. 16, 2021).

12

Standards in Law

Amanda McCormick, University at Buffalo (SUNY)

INTRODUCTION

A good lawyer prepares fastidiously for their practice of law, whether it be as a litigator in the courtroom or as a drafter of contractual agreements. And a good lawyer begins their career in law school. Because standards intersect many areas of legal study, from intellectual property and technology law to business and corporate law, teaching standards literacy to law students is an essential component of today's legal education. As discussed in previous chapters, standards developing organizations (SDOs) operate across industries, and standards are implemented by both private and governmental organizations.

Law schools in the United States have not traditionally focused on standards education as an element of the curriculum, but this is changing with the addition of coursework and the creation of supplemental educational materials by leading institutions, such as the University of Pennsylvania Carey Law School [1–2]. This chapter will provide an overview of the intersection between standards and two areas of U.S. law: *administrative law* and *intellectual property and technology law*.

Administrative Law

Administrative law is the study of governmental agencies, which are found on the federal and state levels. Agencies are delegated powers by Congress or state legislatures that permit them to administer, interpret, and enforce laws [3]. Of particular importance to the discussion is the role of administrative agencies in the rulemaking process. The U.S. rulemaking process, as illustrated in Figure 12.1, is a complicated procedure in which an agency promulgates binding law [4]. Standards developed by SDOs, that is, privately developed technical standards, may be adapted by agencies during the rulemaking process [5]. A standard may not be (and often is not) printed within the final rule; instead, the text of the standard is "incorporated by reference" into the rule. The standard is thus referred to within the final and binding rule as published in the *Code of Federal Regulations*, but the full text of the standard is not reprinted within [5].

This practice raises several issues for law students, scholars, and those in industry: Where is the standard published, if not in the *Code of Federal Regulations*? What is the impact of this practice on those regulated by government agencies? This lack of clarity is not a minor issue; according to the National Institute of Standards (NIST) Standards Incorporated by Reference database (note: unfortunately, not updated since 2016), regulations from the Environmental Protection Agency (EPA) incorporate over 9,000 standards from SDOs, such as 3M Corporation, ASTM International, and the American Society of Mechanical Engineers [6]. The EPA, it should be noted, regulates the agriculture, automotive, construction, electric utilities, oil and gas, and transportation sectors. Where does an interested party find the relevant standard? The answer is not simple.

Suppose that you are a member of the information technology team at a large health care system. Your team uses the Acme Electronic Health Record Platform, and a question has arisen regarding privacy controls in the platform. In order to research this issue, you turn to the U.S. Department of Health and Human Services, which directs you to Title 45 of the *Code of Federal Regulations* (C.F.R.), subtitle A, subchapter D [7]. Subchapter D covers "Health Information Technology." You find the section addressing privacy in 45 C.F.R. 170.205 (o):

§ 170.205 Content exchange standards and implementation specifications for exchanging electronic health information.

The Secretary adopts the following content exchange standards and associated implementation specifications:

U.S. Rulemaking Process

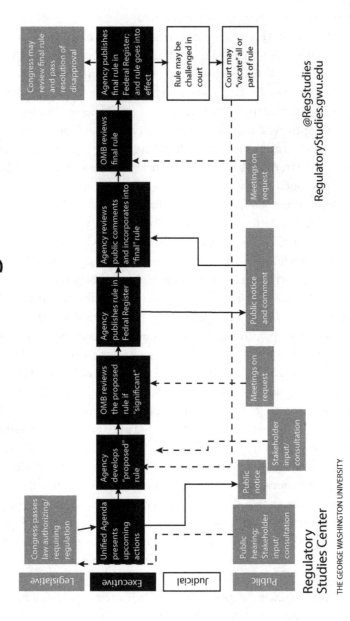

Legislative
- Congress passes law authorizing/requiring regulation
- Congress may review final rule and pass resolution of disapproval

Executive
- Unified Agenda presents upcoming actions
- Agency develops "proposed" rule
- OMB reviews the proposed rule if "significant"
- Agency publishes rule in Federal Register
- Agency reviews public comments and incorporates into "final" rule
- OMB reviews final rule
- Agency publishes final rule in Federal Register; and rule goes into effect

Judicial
- Rule may be challenged in court
- Court may "vacate" all or part of rule

Public
- Public hearing; Stakeholder input/consultation
- Public notice
- Stakeholder input/consultation
- Meetings on request
- Public notice and comment
- Meetings on request

Regulatory
Studies Center

THE GEORGE WASHINGTON UNIVERSITY

@RegStudies
RegulatoryStudies.gwu.edu

FIG. 12.1. U.S. rulemaking process modified from the George Washington University Regulatory Studies Center.

(o) Data segmentation for privacy—
(1) Standard. HL7 Implementation Guide: Data Segmentation for Privacy (DS4P), Release 1 (incorporated by reference in § 170.299).
(2) [Reserved] [8].

Next, as directed, you turn to 45 C.F.R. 170.299, which states:

§ 170.299 Incorporation by reference.
(a) Certain material is incorporated by reference into this subpart with the approval of the Director of the Federal Register under 5 U.S.C. 552(a) and 1 CFR part 51. To enforce any edition other than that specified in this section, the Department of Health and Human Services must publish a document in the Federal Register and the material must be available to the public. All approved material is available for inspection at the U.S. Department of Health and Human Services, Office of the National Coordinator for Health Information Technology, 330 C Street SW., Washington, DC 20201, call ahead to arrange for inspection at 202-690-7151, and is available from the sources listed below. It is also available for inspection at the National Archives and Records Administration (NARA). For information on the availability of this material at NARA, call 202-741-6030 or go to http://www.archives.gov/federal_register/code_of_federal_regulations/ibr_locations.html [9].

Subsequently listed within the section are no less than 15 SDOs, including the American National Standards Institute (ANSI), ASTM International, and the International Telecommunication Office. You then search for the "HL7 Implementation Guide: Data Segmentation for Privacy" referenced in 45 C.F.R. 170.205 (o) [8]. The guide is referenced in 45 C.F.R. 299 (f) (25) and is available by contacting Health Level Seven International (HL7) with address information provided in 45 C.F.R. 299 (f) [8]. You then, as directed, visit the website given for HL7, a company that provides standards to the health care industry's administrative data sector, and type the name of the standard into the search bar [10]. The search results direct you to the appropriate web page, where you can read a summary of the standard and download a zip file of the document [11]. Finally, the needed information is found. This laborious process, unfortunately, is more common than not.

An excellent resource exploring the issues surrounding this topic may be found in Emily Bremer's teaching guide, "Technical Standards meet Administrative Law: A Teaching Guide on Incorporation by Reference" [5].

INTELLECTUAL PROPERTY AND TECHNOLOGY LAW

Intellectual property is a broad term that encompasses patents, copyrights, trademarks, and trade secrets, or as elegantly described by the World Intellectual Property Organization, encompasses "creations of the mind" [12]. An in-depth discussion of intellectual property law may be viewed at Cornell University's Legal Information Institute website [13], but, for the purposes of this discussion, we will cover the intersection of standards with patent law and with copyright law.

Patents

The roots of patent law in the United States began with a proposal by Representative Thomas Jefferson at the Constitutional Convention of 1789. This proposal led to a provision in section 8, article 8 of the Constitution authorizing Congress "to promote the progress of science and useful arts, by securing for limited times to authors and inventors the exclusive right to their respective writings and discoveries." The purpose of patent law is to encourage inventors to produce utilitarian works in exchange for exclusive rights for a limited period, with the goal being the enrichment of the public domain [13, 14].

Statutory requirements for patentable inventions are found in Title 35 of the U.S. Code. There are several requirements that inventions must meet for the Patent and Trademark Office (PTO) to issue a patent (for example, the patent application must explain the utility of the claimed invention). The patent permits the owner to exclude others from making, selling, using, offering for sale, or importing the claimed invention, all for a term of approximately 20 years. Patent terms are dependent on several factors; it is best to consult with an attorney for a precise determination [13, 14].

The patent application process is notoriously complex, time-consuming, and costly. Even after a patent is granted, additional issues may arise in relation to the creation of standards. Attorney Melissa Steinman expertly summarizes the issue in a blog post:

There is a fundamental conflict between broad [intellectual property] rights (the exclusive rights granted to an inventor by patent [. . .]) and the necessity of interoperability in the digital economy. In creating standards, the challenge is to balance (a) the individual ownership rights recognized by patent [. . .] laws; (b) the competition values protected by antitrust laws; and (c) the need for compatibility of competitors' products [15].

To realize financial benefits from the grant of the patent, a patent holder is incentivized to offer access to the patented technology through contractual agreements. If a patent is deemed "essential" to a technology, it is a "standard-essential patent" or "SEP" and may be licensed by SDOs to use in developing standards. The surrounding issues are complex and are explored in the book *Patents and Standards: Practice, Policy, and Enforcement* (please visit the open access fourth chapter, "Standards and Intellectual Property Rights Policies") [16]. An excellent teaching guide on this topic has been created by University of Pennsylvania Professor Cynthia L. Dahl, "When Standards Collide with Intellectual Property: Teaching about Standard Setting Organizations, Technology, and *Microsoft v. Motorola*" [17].

Copyright

Grounded in the U.S. Constitution, copyright law is a form of protection for original works of authorship, including "literary, dramatic, musical, architectural, cartographic, choreographic, pantomimic, pictorial, graphic, sculptural, and audiovisual productions" [18]. For a limited time period, copyright provides the holder with the right to reproduce, to make derivative works, to distribute, to publicly perform, and to display the work. Not all works are protected by the copyright law; some may be aged out of the system or may be protected by a license agreement (e.g., a contract between parties or a Creative Commons license) or belong to a class of items that are not copyrightable (e.g., law or government documents) [19].

Carved out of these rights is the fair use doctrine, which permits selective use of a work in order to, among other things, promote research and scholarship. The fair use doctrine is a checklist of four factors that must be weighed to determine whether a use should be deemed fair. The factors examined are: (1) what is the purpose and the character of the use; (2) what is the nature of the copyrighted work; (3) what is the amount and substantiality of the portion used in relation to the copyrighted work as a whole; and (4) what is the effect of the use on the potential market for, or value of, the copyrighted work. The fair use doctrine is applied by courts on a case-by-case basis [13].

Copyright issues are implicated throughout the standards setting. As summarized by Bremer:

> Copyright law presents at least three issues. The first is the eligibility of standards for copyright protection. . . . The second is whether and under what circumstances a government reproduction of copyrighted work may constitute a fair use. . . . The third issue is whether a standard loses its copyright protection when a government entity adopts that standard as a law or incorporates it by reference into law [5].

Turning to the "incorporation by reference" doctrine, which also was discussed above in the administrative law section, note that there is an inherent tension between public access to state law and private SDOs' claimed copyrights in standards. This tension can be seen in a recent case out of New York federal court. An SDO sued an online publisher for providing access to the full text of state code, including the standards that had been "incorporated by reference" into the code [20]. Recall here that copyright law does not apply to certain classes of work, such as law, which fall into the public domain. In a lengthy opinion exploring many aspects of copyright law and standards, the court stated: "At bottom, the controlling authorities make clear that a private party cannot exercise its copyrights to restrict the public's access to the law" [20]. The matter on the whole, however, is far from resolved as federal court rulings are not binding on other jurisdictions. See, for example, the decision in *American Society for Testing v. Public.Resource.Org, Inc.* [21]. Bremer's teaching guide addresses this issue as well as the issues referenced above [5].

The teaching guides addressed in the above discussion are excellent and are well-suited to discussions in law school classrooms. For a case study that may be used with undergraduates with an interest in law and policy, please view "Case Study #3: Standards in the Law Case Study," found in Part IV of this book.

REFERENCES

1 University of Pennsylvania Carey Law School Centers and Institutes, Penn Program on Regulation, "Voluntary codes and standards." https://www.law.upenn.edu /institutes/ppr/codes-standards (accessed Aug. 19, 2021).

2 C. Coglianese and C. Raschbaum, "Teaching voluntary codes and standards to law students," *Administrative Law Review, 71(2)*, pp. 307–313. http://www.admin istrativelawreview.org/volume-71-2.

3 Cornell Law School Legal Information Institute, "Administrative law." https:// www.law.cornell.edu/wex/administrative_law (accessed Aug. 19, 2021).

4 S. Dudley, *Opportunities for stakeholder participation in US regulation*. Regulatory Studies Center of the George Washington University. https://regulatorystudies .columbian.gwu.edu/sites/g/files/zaxdzs4751/files/downloads/GW%20Reg%20 Studies%20-%20Opportunities%20for%20Stakeholder%20Participation%20 in%20Federal%20Regulation%20-%20SDudley.pdf (accessed Aug. 19, 2021).

5 E. Bremer, "Technical standards meet administrative law: A teaching guide on incorporation by reference," *Administrative Law Review, 71(2)*, 2019, pp. 315–352. https://administrativelawreview.org/volume-71-issue-2.

6 NIST Standards Incorporated by Reference Database. https://sibr.nist.gov (accessed Aug. 12, 2021).

7 National Archives Code of Federal Regulations (eCFR). https://www.ecfr.gov /current/title-45/subtitle-A/subchapter-D/part-170/subpart-D (accessed Dec. 21, 2021).

8 National Archives Code of Federal Regulations (eCFR). https://www.ecfr.gov /current/title-45/subtitle-A/subchapter-D/part-170/subpart-B/section-170.205 (accessed Dec. 21, 2021).

9 National Archives Code of Federal Regulations (eCFR). https://www.ecfr.gov /current/title-45/subtitle-A/subchapter-D/part-170/subpart-B/section-170.299 (accessed Dec. 21, 2021).

10 Health Level Seven International (HL7). https://www.hl7.org (accessed Dec. 21, 2021).

11 HL7 Implementation Guide: Data Segmentation for Privacy, Release 1. https:// www.hl7.org/implement/standards/product_brief.cfm?product_id=354 (accessed Dec. 21, 2021).

12 World Intellectual Property Organization, "What is intellectual property?" https:// www.wipo.int/about-ip/en (accessed Aug. 19, 2021).

13 Cornell Law School Legal Information Institute, "Intellectual property." https:// www.law.cornell.edu/wex/intellectual_property (accessed Aug. 19, 2021).

14 U.S. Patent and Trademark Office, "Patent basics." https://www.uspto.gov /patents/basics (accessed Aug. 19, 2021).

15 M. Steinman, "Legal issues affecting standard-setting: Antitrust and intellectual property." https://www.venable.com/insights/publications/2004/04/legal-issues -affecting-standardsetting-antitrust (accessed Aug. 19, 2021).

16 R. Taffet and P. Harris, "Standards and intellectual property rights policies," in *Patents and standards: Practice, policy, and enforcement*, M. Drapkin, K. Kjel-land, D. Long, R. Taffet, and T. Stadheim, eds., pp. 1–28. Arlington, VA: Amer-ican Intellectual Property Law Association, 2018. https://www.morganlewis.com /-/media/files/publication/outside-publication/article/2019/taffet-standards-and -intellectual-property-rights-policies.pdf.

17 C. Dahl, "When standards collide with intellectual property: Standard setting or-ganizations, technology, and *Microsoft v. Motorola*." https://scholarship.law.upenn. edu/faculty_scholarship/2195 and https://www.law.upenn.edu/institutes/ppr/codes -standards (accessed Aug. 19, 2021).

18 U.S. Copyright Office, "Copyright in general." https://www.copyright.gov/help /faq/faq-general.html (accessed Aug. 19, 2021).

19 A. McCormick, "Copyright, Fair use and the digital age in academic libraries: A review of the literature," *School of Information Student Research Journal*, 4(2), 2014. https://doi.org/10.31979/ 2575-2499.040205

20 International Code Council v. UpCodes, Inc., U.S. District Court, Southern District of New York, 17 Civ. 6261, 2020. https://scholar.google.com/scholar_case?case=1013934 9383067759286&q=international+code+council+v+upcodes&hl=en&as_sdt=6%2 C33&inst=17395704991083290304

21 Stanford Libraries Copyright and Fair Use, "*American society for testing v. Public. resource.org, Inc.*" https://fairuse.stanford.edu/case/american-society-for-testing -v-public-resource-org-inc (accessed Aug. 19, 2021).

13

Standards in Health Sciences

Suzanne Fricke, Washington State University

ighly publicized examples exist of medical device recalls, or of standards not being incorporated in medical products. One example is the early failure of the Affordable Care Act website due to health care plans failing to implement the ASC X12 standards for enrollment [1]. While publications suggest integrating medical devices into the health science curriculum in the form of mobile applications and wearable sensors, few talk about discussing standards of these and other medical devices [2]. While health science students may not require an exhaustive understanding of standards, they must recognize where standards exist or may be needed, value the potential of interoperable systems that use standards, and know-how to report adverse events.

AUDIENCE FOR THIS CHAPTER

This chapter is directed at librarians and faculty who work with health sciences students who need to understand complex systems to improve patient and population health. As well it is directed at librarians and faculty who work with engineering students who need to understand the complexity of interacting standards and clinical contexts in order to create products for the health care environment. Both are part of interdisciplinary teams that require a greater understanding of the medical product life cycle and reporting systems [3–5]. This chapter may also benefit librarians and faculty who work with programs in regulatory science or regulatory affairs that develop new methods and computational tools for assessing safety and risk [6–9]. These regulatory science degree programs are often interdisciplinary and housed within a variety of academic departments including health sciences, biopharmaceutics, engineering, business, law, and environmental science. Also, medical schools and large medical centers are increasingly focusing on innovation and rapid design thinking [10]. As a result, librarians are called to work with entrepreneurial teams developing new drugs and biologicals, health information systems, diagnostic equipment, medical instruments and devices, or health care quality improvement processes [6, 11–15].

While we often assume that students today rapidly learn how to use new products and technologies, this rapid assimilation may not always come with an understanding of the purpose of these tools, or the systems underlying their creation and regulation [16]. As a result of this lack of understanding, medical products are used inconsistently by health care providers, and post-market adverse events are underreported due to time or culture constraints [17]. Students and professionals also may not fully understand the role that standards play in transitions of care and post-market analysis. By graduating health care professionals who lack understanding of these systems, we may inadvertently impact the ability to extract meaningful data about existing products and potentially impair the future creation of safe and effective products. As well, we may not be properly preparing graduates who can assure that engineering standards remain up to date with current understanding of biomechanics, physiology, technology, security, and standards of care [18]. As we look to a future of increased precision medicine and artificial intelligence, adherence to standards will facilitate increased clinical decision support enabled by learning health care systems.

THEORETICAL FRAMEWORK

Health science librarians frequently focus their teaching on the Association of American Medical Colleges (AAMC) Entrustable Professional Activity (EPA) 7 for entering residents, that of evidence-based practice [19]. They assist students and providers in finding, evaluating, and synthesizing the best available evidence (e.g., journal articles, clinical practice guidelines, etc.) for clinical decision making. They may be less comfortable with their role in AAMC EPA 5 documentation, EPA 9 interprofessional teams, and EPA 13 system failures.

Even further, they may fail to look more closely at Accreditation Council for Graduate Medical Education (ACGME) Core Competencies for graduating residents, specifically Core Competency 5, practice-based learning and improvement (PBLI), and 6, systems-based practice, which were added in 1999 and 2002 [20–24]. These competencies were created to acknowledge that health care providers no longer work as solo practitioners, and increasingly need to understand complex systems and work with interdisciplinary team members—nurses, systems administrators, insurance companies, dieticians, social workers, pharmacists, biomedical engineers, and others—to provide care for patients. Previous authors have written about the difficulty in teaching and assessing systems-based practice in health care, even though it has been standard in engineering for decades [25–28]. In a mixed-methods study by Ackerman, et al. of a cardiology outpatient clerkship, students preferred gaining clinical skills through direct client-patient interactions, over systems-based practice objectives focused on workflow, patient user experience, and follow-up communication [25]. Systems-based practice is most frequently addressed through quality improvement exercises such as clinical audits or morbidity and mortality rounds [21, 29–30]. Librarians have mapped these competencies to the ACRL Framework, though high-level documents may fail to translate to logistical examples of teaching for these competencies [31].

Teaching standards to health care professionals presents one opportunity for greater understanding of systems-based practice and practice-based learning and improvement. Systems-based practice requires the learner to adapt to changes in health care and reporting systems [20]. While much is focused on human-centered design and creating and using systems that understand people/user needs, less in health care is focused on the opposite, creating individuals and populations who understand standards underlying these systems and the need for interoperability with other systems [32–34].

HISTORY OF STANDARDS IN HEALTH CARE

In the United States, the Food, Drug, and Cosmetic Act (FDCA) passed in 1938 first gave authority for food and drug safety to the Food and Drug Administration

(FDA). The FDA amended the Food, Drug, and Cosmetic Act in 1976 to include medical devices 201(h) and to define three classes of devices:

Class I—do not require premarket approval

Class II—require premarket notification (FDA 510(k)) and post-market surveillance

Class III—approved by the premarket approval (PMA) process including clinical trials for quality, safety, and effectiveness that are similar to drugs

For drugs and devices marketed prior to the amendment, it required the device manufacturer to undergo the premarket authorization process and prove the safety and efficacy of the device to continue marketing it.

The addition of FDA Medical Device Reporting (MDR) in 1984 required manufacturers to report complaints and incidents to the FDA. In 1990, the Safe Medical Device Act amended the FDCA to require device traceability and added requirements by distributors and health care facilities to report post-market incidents to the FDA. In 1995, reporting forms were standardized and foreign device manufacturers were required to comply with the same regulations, and a 1998 FDCA amendment adjusted the Safe Medical Device Act to require distributors to report complaints only, not incidents. The 2002 Medical Device User Fee and Modernization Act focused on premarket and reprocessed devices. The 2016 21st Century Cures Act improved the regulation of combination products, created procedures for new indications for approved drugs, expedited processes for biologics and medical devices in response to health needs, and set parameters for collecting sustainable real-time post-market safety and adverse reporting data from networked devices. The FDCA was extended in 2017 with the Food and Drug Administration Reauthorization Act (FDARA).

In the United States, the FDA adopts technical, engineering, or information exchange specifications or terminologies developed by national or international standards developing organizations, or other government agencies. These are incorporated into Current Good Manufacturing Practice regulations for the manufacturing of products under the Center for Devices and Radiological Health (CDRH), the Center for Biologics Evaluation and Research (CBER), or the Center for Drug Evaluation and Research (CDER). While veterinary devices and drugs are regulated by the FDA, veterinary biologics fall under the jurisdiction of the USDA Animal Plant Health Inspection Service (APHIS) under 9 CFR E: 101-118.

Oversight standards for interoperability, privacy, and security of networked devices fall under the Department of Health and Human Services Office of the National Coordinator for Health Information Technology (ONC), which maintains www.healthIT.gov.

In Europe, the Global Harmonization Task Force on medical devices formed in 1992 just prior to the creation of the European Union. The International Medical Device Regulators Forum replaced this in 2011. Formal medical device regulation (EU 2017/745) requiring greater post-market follow-up was created in 2017, for application in 2021 [35–36]. Implementation was delayed until spring 2024 as the European Database on Medical Devices (EUDAMED) prepared to register devices and assure unique identifiers.

Several medical device classification/nomenclature systems exist around the world. Some of the more common are the United Nations Standard Products and Services Code (UNSPSC), the Global Medical Device Nomenclature (GMDN), the Universal Medical Device Nomenclature System (UMDNS), the Generic Implant Classification (GIC), and the European Medical Devices Nomenclature (EMDN). Use of a particular system often is decided based on a nomenclatures structure (hierarchy or polyhierarchy), licensing (free or copyright), granularity of description, and use by specific disciplines or partnering organizations. EUDAMED requires the use of EMDN because, unlike the proprietary polyhierarchy UMDNS system, it is a freely available hierarchy [37–38].

CLASSIFICATION OF HEALTH CARE PRODUCTS COVERED BY STANDARDS

Health-related engineering standards cover medical devices, information technology, drugs, biologicals, and facilities. This section will address each of these segments in turn.

The definition of "medical devices" is poorly understood. This term incorporates an array of equipment encountered in diverse settings and disciplines. While the phrase is used frequently in engineering settings, it is rarely encountered in health science curriculums. Medical devices may include laboratory and imaging diagnostic equipment, remote and bedside patient monitors, drug delivery systems, drug manufacturing materials and equipment, medical implants, personal protective equipment, and surgical instruments and robots. Laboratory equipment combines reference (tests and analysis) and metrology (measurement) standards with materials and network capabilities. Previous authors have classified medical devices by their function (therapeutic, diagnostic, and analytical), data type (standard DICOM, HL7, XML, or nonstandard image data), connections to networks, and data flow.

Pharmaceutical drugs, chemical substances that affect physiology or psychology, are regulated separately from biologicals or biologics, originating from living cultures or blood, a category that includes vaccinations, blood products, and a growing array of immunotherapies.

Health information technology (HIT) incorporates electronic health records, information and communications technology (ICT), telehealth, standard file formats unique to health care (such as DICOM radiology images), algorithms, security/privacy, health information exchange (HIE), and a growing number of networked medical devices in what is sometimes referred to as the Internet of Medical Things (IoMT) [39].

Health information technology systems frequently use permanent identifier standards and terminology standards designed to represent the context of the health care setting or injury. Identifier and terminology uptake may vary across countries based on mandates, incentives, and the degree to which health care is publicly administered. Controlled vocabularies that seek to define context can be a good entry point for librarians, and some systems may map clinical terminology to educational objectives, or to controlled vocabularies used to index literature, such as the National Center for Biotechnology Information (NCBI) Medical Subject Headings (MeSH).

While the focus of this book is on technical/engineering standards, the FDA recognizes a growing number of combination products, and these products often require consulting multiple categories of standards. For example, human drugs and biologics are regulated by the FDA, and their manufacturing, packaging, and delivery systems are subject to materials and manufacturing standards. Tissue-engineered medical products (TEMPs) used as implants in regenerative medicine are composed of both biological and synthetic materials. Health care facilities that are subject to standards for air quality, water, waste, materials, energy, design, and networks that impact patient safety, may also choose to pursue the Leadership in Energy and Environmental Design (LEED) standard for health care, or install SMART operating rooms. Environmental standards for water, air, and waste may force health care providers to find new methods for necessary tasks such as cleaning and sterilization. Standards for data applicable to HIT, ICT, and HIE can assure that data generated by micro-electromechanical systems (MEMS), which incorporate mechanical and networked electronic elements, are compatible with other systems. At the same time, data generated by these products need to meet privacy and security standards. Standards must be compatible with multiple organizational standards and medical practice guidelines, which directly impact diagnostics or patient treatment. For instance, ISO medical laboratory quality standards are compatible with the Laboratory Medicine Practice Guidelines (LMPGs) from the American Association of Clinical Chemistry, which conforms to the National Academy of Medicine Committee on Standards.

HEALTH CARE ENVIRONMENT

The health care environment is unique in many ways. Health care systems are complex. Providing care involves a variety of interdisciplinary stakeholders, multiple systems, and frequent transitions of care [32]. Patients themselves are members of their own health care team. Highly trained professionals use health care devices and systems. In many cases, the outcome of their use of medical devices may be dependent on skill and technique, while other times their use of devices (such as health information systems) is considered secondary to their main job [40].

Health care accounts for the highest number of professional malpractice claims. Patient safety itself has standards, and it is safety that becomes the number one goal driving the use of standards [41]. While studies show that implementing HIT does reduce medication errors and improves compliance, the overall impact on patient safety requires further study [42]. Furthermore, in addition to inherent medical and surgical risks associated with health care, the expectation for security and privacy standards is high due to the Health Insurance Portability and Accountability Act (HIPAA). Health care providers are at risk of security breaches and ransomware attacks.

Health care increasingly is moving away from clinic-centered care to a continuum of care emphasizing prevention—and intervention when risk is determined [43]. While regulatory agencies and manufacturers have traditionally used a system of premarket clinical trials and postmarket reaction to problems after reported incidents, the potential exists through machine learning for more dynamic risk assessment relying on Markov modeling. This can facilitate real-time decision support and earlier intervention [44].

As health care becomes more automated, medical education runs the risk of focusing too much on preparing physicians to work within a system, and not enough on preparing students to change the system when it fails to advance health equity [24]. As a result, health education is now focusing more on personalized medicine, and this starts with recruiting diverse participants for medical device and drug clinical trials [45].

Products often interact with other products and with the human body, through either physical contact, chemical interaction, or technical connection. Physical contact requires strict standards for withstanding and assuring sterility. Chemical contact between products requires preventing incompatibility. Technical connections require interoperability with a complex health system that includes monitors, medical records, financial systems, insurance claims, patient portals, and quality improvement systems. As well, alarms or alerts designed to increase patient safety may inadvertently cause harm if they contribute to provider fatigue in the health care environment.

EDUCATION OF HEALTH CARE PROVIDERS

Previous authors have emphasized the importance of case-based experiential learning for regulatory science [6]. While case-based learning is common in medicine, it may be less commonly seen in relation to medical product development and product use in health care settings. For health science professionals, understanding regulatory science begins with an understanding of evidence-based practice, critical appraisal, and experience applying real-time data to patient care. A modern example of the important role of standards exists within the current world problem of antibiotic resistance. Health care providers should consult Clinical and Laboratory Standards Institute (CLSI) standards for species to accurately correlate in vitro culture and sensitivity results with patient clinical parameters in order to select antibiotic protocols that prevent antibiotic resistance [46]. Once students have an advanced understanding of evidence levels, and the application of population-level evidence to clinical practice, they are prepared to understand how their own documentation, using standard calculations, file formats and terminologies in health information systems, impacts the creation of future practice-based evidence or real-world evidence (RWE) [47–49].

A step beyond evidence-based medicine involves teaching students quality improvement methods that dive deep to the level of consulting and assessing current standards. These are best taught in interdisciplinary team settings, reflective of the work environment. Students should identify the presence of post-market reporting systems and understand that health care providers are not powerless to change unsafe and ineffective products; however, they need outcomes data, and they need to collect it in a standard way. Because many government and manufacturer reporting systems have not been transparent or consistent across countries or states in the past, and because health care providers do not have access to device and operator-specific information from scientific studies, many health care providers have become proponents of international medical device registries, which provide data independent of industry [18, 50–51]. Only through regular use of reporting systems and registries can health care providers accurately identify where adverse events are associated with standards or variability in standards, and not medical error or patient factors [18]. As Rome states, "Clinician and patient engagement in post-market surveillance and comparative effectiveness research remains imperative" [52].

Adverse reporting databases in the United States include the FDA Adverse Event Reporting System (FAERS) Public Dashboard, which displays human drug and biologic adverse reports since 1968 by region, report type, seriousness of report, type of reporter (health care provider vs. consumer), age, and sex [53]. The database is updated weekly and searchable by generic and proprietary product names. The FDA also maintains the Medical Device Recall Database and the Manufacturer and User Facility Device Experience (MAUDE) Database of Medical Device Reports (MDRs) submitted to the FDA by mandatory reporters (manufacturers, importers, and device user facilities) and voluntary reporters such as health care professionals, patients, and consumers [54]. The FDA Sentinel Initiative provides training and data from partner institutions' electronic health and billing systems to evaluate post-market drug and biologic safety. Working with data created from electronic health records and reporting systems informs how students document in health information systems in the future [55].

Health science students should also practice with medical terminologies/ontologies used in HIT systems so that they understand their power to collocate like cases and enable collective data. The use of virtual patients in electronic health record simulation systems helps to make these ontologies transparent through structured data input and drop-down menus, in place of free text fields. Teaching with the use of actual or simulated hospital electronic health record systems can make these standards even more transparent to users and encourage health professionals to be involved in the ongoing development of these terminologies. Beyond learning medical terminology, the use of ontologies helps to define complex relationships and contexts. For example, the recent update to ICD-10 diagnostic codes have received mixed reviews due to their insistence on exact descriptions. While this can be frustrating for practitioners, it can provide an interesting exercise for students. Organizations regularly release humorous lists of ICD-10 codes, such as W56.01 "bitten by dolphin" and Y93.D "arts and crafts injury." While engaging, these codes also serve to make students consider who might want to collocate this data. Students can use online terminology browsers or metathesauri to identify codes.

Exposure to health care environments, or simulated health care environments, helps product developers understand time constraints, cognitive load, tissue and chemical exposure, anatomic barriers, and other limitations that arise in certain settings where a product may be used. Existing health science simulation laboratories are underutilized as learning environments for broader entrepreneurial groups. When products are complex combination products, such as drug delivery systems or smart wearables, relevant standards bridge multiple standards developing organizations, and they may be best identified by interdisciplinary teams working in simulated settings.

As well, incorporating case scenarios with remote networked devices and transitions of care to other health care facilities will help teams understand the need for interoperable standards like Fast Health Care Interoperability Resources (FHIR) in increasingly complex health information networks. At the same time, teams may recognize the limitations of proprietary systems in health information exchange, and the potential barrier that proprietary data can be to patient safety.

As devices become more networked, and the potential for computational modeling, real-time data dashboards, and RWE increases, we are moving from a reliance on post-market reporting systems to point-of-care risk mitigation. We can prepare students for this by providing opportunities to interpret adverse event databases, crowd-sourced datasets, and data generated by personal devices. Future jobs will require professionals to use these data for risk management at the population, patient, and individual device level. Health care professionals aided by quantitative decision support tools that weigh multiple factors need skills in applying data to patient care, assessing risk management tools for accuracy and bias, and communicating risk effectively to the public.

RESOURCES FOR TEACHING

TYPES OF MEDICAL PRODUCTS

This simplified hierarchy defines the main medical products encountered in teaching health care standards.

1. Diagnostic equipment
   a. Laboratory
   b. Imaging
2. Monitors
   a. Remote sensors/wearables
   b. Hospital
3. Drug delivery systems
4. Drug manufacturing and packaging
5. Medical implants
6. Personal protective equipment (PPE)
7. Instruments
8. Informatics/health information technology
   a. Electronic health record (EHR)
   b. Telehealth
   c. File formats for health information exchange (HL7, DICOM, XML)
   d. Algorithms
   e. Metadata
9. Drugs
10. Biologics
11. Robotics
12. Facilities
13. Sterilization

COMMON STANDARDS USED IN HEALTH CARE

Engineering or Technical Standards [56,57]

ISO (International Organization for Standardization)
- ISO 10993 chemical characterization of medical devices
- ISO 10993-x biological evaluation of medical devices; biocompatibility, genotoxicity
- ISO/IEEE 11073 medical device communication standards applicable to open EHR
- ISO 13485 quality management applicable to medical devices; including OEMs (original equipment manufacturer task-oriented parts such as pressure sensing)
- ISO 14001 environmental compliance
- ISO 14155 clinical trials
- ISO 14971 risk management
- ISO 11784 or 11785 animal radio-frequency identification microchip [59]
- ISO/TS 21526 health informatics/metadata
- ISO 27001 cybersecurity
- ISO/IEC 29119 software [56]
- ISO 500001 energy efficiency
- ISO 90001 quality management

IEC (International Electrotechnical Commission)
- IEC 60601 a series of standards applicable to medical electrical equipment (relevant to DICOM)
- IEC 62304 medical device software (relevant to DICOM)
- IEC 62366-1 medical device usability
- IEC 80001 risk management for IT networks
- IEC 82304-1 product safety [44]

ASME (American Society of Mechanical Engineers)—International verification and validation of computational modeling

ASTM International (formerly American Society for Testing and Materials)—Standards for packaging, health informatics, pharmaceuticals, materials including biomaterials, devices, anesthetic equipment, respirators, and emergency services.

NIST (National Institute of Standards and Technology)—Best practices focused on risk assessment[39], measurement standards; relation to HIT FHIR interoperability

NISO (National Information Standards Organization)—U.S. information standards for publishing, bibliographic, and library applications

- Z39—for incorporating information standards into products (standard used for the Infobutton in electronic health records)

IEEE 1012 verification and validation [56]

CLSI (Clinical and Laboratory Standards Institute)—Medical laboratory testing

AIUM (American Institute of Ultrasound in Medicine)—Safety, maintenance, and calibration of ultrasound equipment, records, and personnel

NEMA (National Electrical Manufacturers Association)—Electrical and medical imaging standards

Terminologies

- ICD (International Classification of Diseases)—World Health Organization codes for diseases, interventions, and function/disability/environments
- IHTSDO (International Health Terminology Standards)
  - SNOMED-CT (Systematized Nomenclature of Medicine—Clinical Terms) international standard for electronic health records provides machine readable terminology covering clinical findings, symptoms, diagnoses, procedures, body structures, organisms and other etiologies, substances, pharmaceuticals, devices, and specimens
- LOINC (Logical Observation Identifiers Names and Codes)—U.S. medical laboratory terminology standard
- RxNorm—U.S. pharmaceutical name standard
- CAS RN (Chemical Abstract Service Registry Number)—permanent identifier for inorganic and organic compounds, minerals, and alloys
- NCBI RefSeq (National Center for Biotechnology Information Reference Sequence)—publicly available DNA and RNA sequences
- CPT (Current Procedural Terminology)—American Medical Association codes for medical services
- CLSI (Clinical and Laboratory Standards Institute) Harmonized Terminology Database
- UNSPSC (United Nations Standard Products and Services Code)—international freely available hierarchy for devices not specific to medicine

- UMDNS (Universal Medical Device Nomenclature System)—international proprietary polyhierarchy for medical devices accepted by the World Health Organization and produced by the Emergency Care Research Institute (ECRI)
- GIC (Generic Implant Classification)—proprietary hierarchy developed specifically for orthopedic devices
- GMDN (Global Medical Device Nomenclature)—international freely available polyhierarchy for medical devices.
- EMDN (European Medical Device Nomenclature)—international freely available hierarchy for medical devices.

Terminology/Ontology Browsers
- IHTSDO SNOMED Browser https://browser.ihtsdotools.org
- UMLS Metathesaurus Browser includes names and codes from standard biomedical vocabularies including SNOMED CT, RxNorm, LOINC, MeSH, CPT, ICD-10-CM, MedDRA, Human Phenotype Ontology, and more https://uts.nlm.nih.gov/uts/umls/home

Information Exchange
- HL7—international standards for health information exchange/sharing/interoperability
  - Version 2—syntactics
  - Version 3—semantics
- Fast Health Care Interoperability Resources (FHIR)
- Clinical Document Architecture (CDA) for documents
- DICOM (Digital Imaging and Communications in Medicine) international radiology standards for transmitting, storing, retrieving, printing, processing, and displaying medical images
- ASC (Accredited Standards Committee) X12 national standards for electronic data exchange (EDI) for insurance enrollment and billing

Safety
- OSHA (U.S. Deptartment of Labor Occupational Safety and Health Administration) occupational health and safety standards
- IAEA (International Atomic Energy Agency) standards for medical exposure to radiation

Privacy

- HIPAA (Health Insurance Portability and Accountability Act of 1996) protects individually identifiable health information in the United States
- PIPEDA (Personal Information Protection and Electronic Documents Act) protects individually identifiable health information in Canada
- State and province level consumer privacy
- EU GDRP and U.K. GDRP (General Data Protection Regulation)

Pharmaceutical Reference Standards

USP (United States Pharmacopeial Convention)—documentation and reference standards for nonproprietary naming, packaging, biostrength, quality and purity of drugs, biologics, medical devices, and dietary supplements.

Several efforts are underway to bring disparate data from research, clinical, and billing systems into a common data format using harmonized terminologies [58]. The largest such effort is the Observational Medical Outcomes Partnership (OMOP) Common Data Model (CDM) working with Observational Health Data Sciences and Informatics (OHDSI) partners developing open-source tools [60]. This effort will integrate with FHIR, and with U.S. government data standards as demonstrated in the National COVID Cohort Collaborative [61].

Government

- U.S. Food and Drug Administration, https://www.fda.gov—the FDA website is organized by information about products, topics, and information directed at specific groups of users.
- FDA Office of Combination Products Guidance Documents, https://www.fda .gov/about-fda/office-clinical-policy-and-programs/office-combination -products
- Health Canada List of Recognized Standards for Medical Devices, https:// www.canada.ca/en/health-canada/services/drugs-health-products/medical -devices/standards/list-recognized-standards-medical-devices-guidance.html
- European Commission European Database on Medical Devices (EUDAMED), https://ec.europa.eu/tools/eudamed
- European Medicines Agency, https://www.ema.europa.eu/en—European Public Assessment Reports (EPARs) for human, veterinary, and herbal medications

Search for Standards

- FDA Consensus Standards, https://www.accessdata.fda.gov/scripts/cdrh /cfdocs/cfstandards/search.cfm
- ASTM Medical Device Standards by category, https://www.astm.org/Stan dards/medical-device-and-implant-standards.html
- USP Reference Standards, https://www.usp.org/reference-standards

Adverse Event Databases

- FDA Adverse Event Reporting System (FAERS) Public Dashboard displays human drug and biologic adverse reports since 1968 by region, report type, seriousness of report, type of reporter (health care provider vs. consumer), age, and sex. The database is updated weekly and searchable by generic and proprietary product name. https://www.fda.gov/drugs/questions-and -answers-fdas-adverse-event-reporting-system-faers/fda-adverse-event-reporting -system-faers-public-dashboard
- MAUDE—Manufacturer and User Facility Device Experience Database of Medical Device Reports (MDRs) submitted to the FDA by mandatory reporters (manufacturers, importers, and device user facilities) and voluntary reporters such as health care professionals, patients, and consumers. https://www .accessdata.fda.gov/scripts/cdrh/cfdocs/cfmaude/search.cfm
- FDA Medical Device Recalls Database, https://www.accessdata.fda.gov /scripts/cdrh/cfdocs/cfRES/res.cfm
- FDA Sentinel Initiative, https://www.sentinelinitiative.org—provides training and data from partner institutions' electronic health and billing systems to evaluate post-market drug and biologic safety

Adverse Event Reporting

- FDA Safety Reporting Portal, https://www.safetyreporting.hhs.gov
- MedWatch—FDA's medical product safety reporting portal for health professionals, patients, and consumers on prescription medicines, over-the-counter medicines, non-vaccine biologicals, and medical devices, https://www.access data.fda.gov/scripts/medwatch
- MedWatchLearn resource for teaching students, health professionals, and consumers on how to report problems to the FDA, https://www.accessdata .fda.gov/scripts/MedWatchLearn
- Vaccine Adverse Event Reporting System (VAERS), https://vaers.hhs.gov

- Animal Drug and Device Reporting system, https://www.fda.gov/animal
  -veterinary/report-problem/how-report-animal-drug-and-device-side-effects
  -and-product-problems
- Animal adverse event reporting for manufacturer, https://www.fda.gov/animal
  -veterinary/report-problem/veterinary-adverse-event-reporting-manufacturers

Organizations
- International Medical Device Regulators Forum (IMDRF), https://www.imdrf.org
- National Institute for Innovation in Manufacturing Biopharmaceuticals (NIIMBL), https://niimbl.force.com
- ECRI (originally Emergency Care Research Institute)—international nonprofit safety organization maintains the Universal Medical Device Nomenclature System (UMDNS)

REFERENCES

1 E. Hu, "All things considered. Number of the year: How an obscure government code—834—became big in 2013," *National Public Radio*. https://www.npr .org/sections/alltechconsidered/2013/12/20/255834823/how-an-obscure-government -code-834-became-big-in-2013

2 T. D. Aungst and R. Patel, "Integrating digital health into the curriculum—Considerations on the current landscape and future developments," *Journal of Medical Education and Curricular Development, 7*, Jan/Dec 2020, p. 2382120519901275. https://doi.org/10.1177/2382120519901275.

3 Y. Yazdi, "Developing innovative clinicians and biomedical engineers: A case study," *American Journal of Preventive Medicine, 44*(1), Jan. 2013, pp. S48–50. https://doi.org/10.1016/j.amepre.2012.09.013.

4 A. Thakur, S. Soklaridis, A. Crawford, B. Mulsant, and S. Sockalingam, "Using rapid design thinking to overcome COVID-19 challenges in medical education," *Academic Medicine, 96*(1), Jan. 2021, pp. 56–61. https://doi.org/10.1097 /acm.0000000000003718.

5 P. E. Patterson, "A systematic approach for introducing innovative product design in courses with engineering and nonengineering students," *Biomedical Sciences Instrumentation, 43*, 2007, pp. 91–94.

6 J. E. Adamo, E. E. Wilhelm, and S. J. Steele, "Advancing a vision for regulatory

science training," *Clinical and Translational Science, 8*(5), Oct. 2015, pp. 615–618. https://doi.org/10.1111/cts.12298.

7 T. M. Morrison, P. Pathmanathan, M. Adwan, and E. Margerrison, "Advancing regulatory science with computational modeling for medical devices at the FDA's Office of Science and Engineering Laboratories," *Frontiers in Medicine, 5*, Sep. 2018. https://doi.org/10.3389/fmed.2018.00241.

8 P. Spindler, K. F. Bach, M. Schmiegelow, N. Bedlington, and H.-G. Eichler, "Innovation of medical products: The evolution of regulatory science, research, and education," *Therapeutic Innovation & Regulatory Science, 50*(1), 2016, pp. 44–48.

9 D. Drago, S. Shire, and O. Ekmekci, "Improving regulatory education: Can we reconcile employers' expectations with academic offerings?," *Therapeutic Innovation & Regulatory Science, 50*(3), 2016, pp. 330–336.

10 W. Liu, J. Yang, and K. Bi, "Factors influencing private hospitals' participation in the innovation of biomedical engineering industry: A perspective of evolutionary game theory," *International Journal of Environmental Research and Public Health, 17*(20), Oct. 13, 2020. https://doi.org/10.3390/ijerph17207442.

11 N. Shin, K. Vela, and K. Evans, "The research role of the librarian at a community health hackathon—a technical report," *Journal of Medical Systems, 44*(2), Dec. 18, 2019, p. 36. https://doi.org/10.1007/s10916-019-1516-x.

12 A. Singh, D. Ferry, and S. Mills, "Improving biomedical engineering education through continuity in adaptive, experiential, and interdisciplinary learning environments," *Journal of Biomechanical Engineering, 140*(8), Aug. 1, 2018, pp. 0810091-8. https://doi.org/10.1115/1.4040359.

13 A. W. Eberhardt, O. L. Johnson, W. B. Kirkland, J. H. Dobbs, and L. G. Moradi, "Team-based development of medical devices: An engineering-business collaborative," *Journal of Biomechanical Engineering, 138*(7), Jul. 1, 2016, pp. 0708031-5. https://doi.org/10.1115/1.4032805.

14 A. J. Carroll, S. J. Hallman, K. A. Umstead, J. McCall, and A. J. DiMeo, "Using information literacy to teach medical entrepreneurship and health care economics," *Journal of the Medical Library Association, 107*(2), Apr. 2019, pp. 163–171. https://doi.org/10.5195/jmla.2019.577.

15 C. Rika Wright, Z. Shamika, and S. Arif, "Project-based learning: Engaging biomedical engineering sophomores through a collaborative vein-finder device project with nursing," Salt Lake City, Utah, June 23, 2018. https://peer.asee.org/30903.

16 N. Selwyn, "The digital native—myth and reality," in *Aslib Proceeding*. Bradford: Emerald Group Publishing Limited, 2009.

17 M. L. L. Leape, *Making healthcare safe: The story of the patient safety movement.*

Cham, Switzerland: Springer, 2021. https://link.springer.com/book/10.1007%2F978-3-030-71123-8

18 F. Henshaw et al., "Engineering standards for trauma and orthopaedic implants worldwide: A systematic review protocol," *BMJ Open, 8*(10), Oct. 18, 2018, p. e021650. https://doi.org/10.1136/bmjopen-2018-021650.

19 P. Cocks, W. Cutrer, K. Esposito, and C. Lupi, "EPA 7 schematic: Form clinical questions and retrieve evidence to advance patient care," in *Core Entrustable Professional Activities for Entering Residency*, V. Obeso, D. Brown, and C. Phillipi, eds. Washington DC: Association of American Medical Colleges, 2017.

20 A. G. Lee et al., "Teaching and assessing systems-based competency in ophthalmology residency training programs," *Survey of Ophthalmology, 52*(6), Nov/Dec 2007, pp. 680–689. https://doi.org/10.1016/j.survophthal.2007.08.021.

21 A. M. Tomolo, R. H. Lawrence, and D. C. Aron, "A case study of translating ACGME practice-based learning and improvement requirements into reality: Systems quality improvement projects as the key component to a comprehensive curriculum," *Postgraduate Medical Journal, 85*(1008), Oct. 2009, pp. 530–537. https://doi.org/10.1136/qshc.2007.024729.

22 S. Guralnick, E. Fondahn, A. Amin, and E. A. Bittner, "Systems-based practice: Time to finally adopt the orphan competency," *Journal of Graduate Medical Education, 13*(2), Apr. 2021, pp. 96–101. https://doi.org/10.4300/jgme-d-20-00839.1.

23 J. D. Gonzalo, C. H. Chuang, S. A. Glod, B. McGillen, R. Munyon, and D. R. Wolpaw, "General internists as change agents: Opportunities and barriers to leadership in health systems and medical education transformation," *Journal of General Internal Medicine, 35*(6), Jun. 2020, pp. 1865–1869. https://doi.org/10.1007/s11606-019-05611-5.

24 E. G. Castillo, J. Isom, K. L. DeBonis, A. Jordan, J. T. Braslow, and R. Rohrbaugh, "Reconsidering systems-based practice: Advancing structural competency, health equity, and social responsibility in graduate medical education," *Academic Medicine, 95*(12), Dec. 2020, pp. 1817–1822. https://doi.org/10.1097/acm.0000000000003559.

25 S. L. Ackerman et al., "The action research program: Experiential learning in systems-based practice for first-year medical students," *Teaching and Learning in Medicine, 28*(2), 2016, pp. 183–191. https://doi.org/10.1080/10401334.2016.1146606.

26 M. A. Dolansky, S. M. Moore, P. A. Palmieri, and M. K. Singh, "Development and validation of the systems thinking scale," *Journal of General Internal Medicine, 35*(8), Aug. 2020, pp. 2314–2320. https://doi.org/10.1007/s11606-020-05830-1.

27 M. M. Plack et al., "Systems thinking and systems-based practice across the health professions: An inquiry into definitions, teaching practices, and assessment,"

Teaching and Learning in Medicine, 30(3), Jul/Sep 2018, pp. 242–254. https://doi.org /10.1080/10401334.2017.1398654.

28 P. B. Batalden and D. C. Leach, "Sharpening the focus on systems-based practice," *Journal of Graduate Medical Education, 1*(1), Sep. 2009, pp. 1–3. https://doi .org/10.4300/01.01.0001.

29 P. Varkey, S. Karlapudi, S. Rose, R. Nelson, and M. Warner, "A systems approach for implementing practice-based learning and improvement and systems-based practice in graduate medical education," *Academic Medicine, 84*(3), Mar. 2009, pp. 335–339. https://doi.org/10.1097/ACM.0b013e31819731fb.

30 J. K. Johnson, S. H. Miller, and S. D. Horowitz, "Advances in patient safety. Systems-based practice: Improving the safety and quality of patient care by recognizing and improving the systems in which we work," in *Advances in Patient Safety: New Directions and Alternative Approaches, vol. 2,* K. Henriksen, J. B. Battles, M. A. Keyes, and M. L. Grady, eds. Rockville, MD: Agency for Healthcare Research and Quality (U.S.), 2008.

31 E. A. Brennan, R. S. Ogawa, K. Thormodson, and M. von Isenburg, "Introducing a health information literacy competencies map: connecting the Association of American Medical Colleges Core Entrustable Professional Activities and Accreditation Council for Graduate Medical Education Common Program Requirements to the Association of College & Research Libraries Framework," *Journal of the Medical Library Association, 108*(3), Jul. 1, 2020, pp. 420–427. https://doi .org/10.5195/jmla.2020.645.

32 G. M. Samaras and R. L. Horst, "A systems engineering perspective on the human-centered design of health information systems," *Journal of Biomedical Informatics, 38*(1), Feb. 2005, pp. 61–74. https://doi.org/10.1016/j.jbi.2004.11.013.

33 J. E. McLaughlin, M. D. Wolcott, D. Hubbard, K. Umstead, and T. R. Rider, "A qualitative review of the design thinking framework in health professions education," *BMC Medical Education, 19*(1), Apr. 4, 2019, p. 98. https://doi.org/10.1186 /s12909-019-1528-8.

34 B. Holtz, K. Vasold, S. Cotten, M. Mackert, and M. Zhang, "Health care provider perceptions of consumer-grade devices and apps for tracking health: A pilot study," *JMIR mHealth and uHealth, 7*(1), Jan. 22, 2019, p. e9929. https://doi.org/10.2196 /mhealth.9929.

35 P. Marešová, B. Klímová, J. Honegr, K. Kuča, W. N. H. Ibrahim, and A. Selamat, "Medical device development process, and associated risks and legislative aspects-systematic review," *Frontiers in Public Health, 8*(308), July30, 2020. https:// doi.org/10.3389/fpubh.2020.00308.

36 International Medical Device Regulators Forum, "International Medical Device Regulators Forum Strategic Plan 2021–2025," 2020. https://www.imdrf.org

37 European Commission, "EUDAMED Database." https://ec.europa.eu/tools/euda med

38 World Health Organization, "Standardization of medical devices nomenclature: International classification, coding and nomenclature of medical devices EB150/14 Add. 1." https://apps.who.int/gb/ebwha/pdf_files/EB150/B150_14Add1-en.pdf

39 T. Yaqoob, H. Abbas, and N. Shafqat, "Integrated Safety, Security, and Privacy (ISSP) Risk Assessment Framework for medical devices," *IEEE Journal of Biomedical and Health Informatics, 24*(6), Jun. 2020, pp. 1752–1761. https://doi.org/10.1109/JBHI.2019.2952906.

40 R. E. Hoyt, W. R. Hersh, and E. V. Bernstam, *Health informatics: Practical guide.* Author: Lulu.com, 2018.

41 T. Miclăuş et al., "Impact of design on medical device safety," *Therapeutic Innovation and Regulatory Science, 54*(4), Jul. 2020, pp. 839–849. https://doi.org/10.1007/s43441-019-00022-4.

42 Y. K. Alotaibi and F. Federico, "The impact of health information technology on patient safety," *Saudi Medical Journal, 38*(12), 2017, p. 1173.

43 D.-W. Kim, J.-Y. Choi, and K.-H. Han, "Medical device safety management using cybersecurity risk analysis," *IEEE Access, 8,* 2020, pp. 115370–115382. https://doi.org/10.1109/ACCESS.2020.3003032.

44 S. R. Fox-Rawlings, L. B. Gottschalk, L. A. Doamekpor, and D. M. Zuckerman, "Diversity in medical device clinical trials: Do we know what works for which patients?: Diversity in medical device clinical trials," *The Milbank Quarterly, 96*(3), 2018, pp. 499–529. https://doi.org/10.1111/1468-0009.12344.

45 Clinical and Laboratory Standards Institute (CLSI), *Understanding susceptibility test data as a component of antimicrobial stewardship in veterinary settings,* 2019. https://clsi.org/standards/products/veterinary-medicine/documents/vet09

46 L. W. Green, "Public health asks of systems science: to advance our evidence-based practice, can you help us get more practice-based evidence?," *The American Journal of Public Health, 96*(3), Mar. 2006, pp. 406–409. https://doi.org/10.2105/AJPH.2005.066035.

47 United States Food and Drug Administration, "Real-world evidence." https://www.fda.gov/science-research/science-and-research-special-topics/real-world-evidence

48 European Medicines Agency, "Data Analysis and Real World Interrogation Network (DARWIN EU)." https://www.ema.europa.eu/en/about-us/how-we-work

/big-data/data-analysis-real-world-interrogation-network-darwin-eu

49 G. Raman et al., "AHRQ methods for effective health care," in *Quality of Reporting in Systematic Reviews of Implantable Medical Devices*. Rockville, MD: Agency for Healthcare Research and Quality, 2012.

50 R. E. Gliklich, N. A. Dreyer, M. B. Leavy, "Registries for evaluating patient outcomes: A user's guide." Washington, D.C.: Agency for Healthcare Research and Quality, 2014.

51 B. N. Rome, D. B. Kramer, and A. S. Kesselheim, "Approval of high-risk medical devices in the US: implications for clinical cardiology," *Current Cardiology Reports*, 16(6), 2014, p. 489. https://doi.org/10.1007/s11886-014-0489-0.

52 United States Food and Drug Administration, "FDA Adverse Event Reporting System (FAERS) Public Dashboard." https://www.fda.gov/drugs/questions-and-answers-fdas-adverse-event-reporting-system-faers/fda-adverse-event-reporting-system-faers-public-dashboard

53 United States Food and Drug Administration, "MAUDE—Manufacturer and User Facility Device Experience." https://www.accessdata.fda.gov/scripts/cdrh/cfdocs/cfmaude/search.cfm

54 D. Church and S. Budsberg, "Plenary topic: Can practice-based evidence complement and promote EBVM?," presented at the *Veterinary Evidence Today*, Edinburgh, Scotland, 2016. https://doi.org/10.18849/ve.v1i4.80.

55 A. Coronato, "Engineering high quality medical software: Regulations, standards, methodologies and tools for certification," in *The Institution of Engineering and Technology*, London, United Kingdom, 2018.

56 A. Coronato and A. Cuzzocrea, "An innovative risk assessment methodology for medical information systems," *IEEE Transactions on Knowledge and Data Engineering*, 2020, pp. 1–1. https://doi.org/10.1109/TKDE.2020.3023553.

57 World Small Animal Veterinary Association (WSAVA), *Microchipping—The importance of ISO*, 2020. https://wsava.org/wp-content/uploads/2020/01/Microchipping-The-Importance-of-ISO.pdf

58 E. K. Burrows, H. Razzaghi, L. Utidjian, and L. C. Bailey, "Standardizing clinical diagnoses: Evaluating alternate terminology selection," *AMIA Joint Summits on Translational Science Proceedings*, 2020, pp. 71–79.

59 Observational Health Data Sciences and Informatics (OHDSI), *The Book of OHDSI*. San Bernardino, CA: OHDSI, 2019. https://ohdsi.github.io/TheBookOfOhdsi

60 D. Raths, "HL7, OHDSI to collaborate on single common data model: Data standards organizations will integrate FHIR, OMOP models to create a sin-

gle source for the sharing and tracking of data," Mar. 1, 2021. https://www.hci
nnovationgroup.com/interoperability-hie/standards/news/21212141/hl7-ohdsi-to
-collaborate-on-single-common-data-model

PART IV

Case Studies

This section provides a variety of case studies collected from librarians, instructors, and faculty that integrate standards into the classroom, lab, learning management software, or project-based learning. The case studies in this section span subjects from biomedical engineering to business, and more.

Each case study includes the following elements: a synopsis of the case, information literacy learning outcomes, target audience, description of instruction, description of instructional materials, and assessment of learning. Using the information included in the case studies, the authors hope to inspire others to incorporate standards into their courses or projects and will continue the conversation of standards in the curriculum.

Case Study #1: Mechanical Engineering, *Erin Thomas, Iowa State University*
Case Study #2: First-Year Engineering, *Katie Harding, McMaster University*
Case Study #3: Law, *Amanda McCormick, University at Buffalo*
Case Study #4: Health Sciences, *Suzanne Fricke, Washington State University*
Case Study #5: Business Management, *Margaret Phillips, Heather Howard, Annette Bochenek, and Zoeanna Mayhook, Purdue University*
Case Study #6: Biomedical Engineering, *Joanna Thielen and Jamie Niehof, University of Michigan*
Case Study #7: Civil Engineering, *Xiaowei Wang and Yue Li, Case Western Reserve University*

Case Study #8: Electrical Engineering, *Seyed Hossein Miri Lavasani, Case Western Reserve University*

Case Study #9: Fire Science, *Ya-Ting Liao and Daniela Solomon, Case Western Reserve University*

Case Study #10: Transportation Engineering, *Thomas Abdallah, Metropolitan Transportation Authority Construction & Development, New York City Transit; Yekaterina Aglitsky, Metropolitan Transportation Authority Construction & Development, New York City Transit; Shirley Chen, New York City Mayor's Office of Environmental Remediation; Maria Cogliando, New York University; Louiza Molohides, Columbia University; and Angelo Lampousis, Department of Earth and Atmospheric Sciences, City College of New York, City University of New York*

Case Study #11: Mechanical and Aerospace Engineering, *Daniela Solomon and Ya-Ting Liao, Case Western Reserve University*

Case Study #12: STEM Communication/Technical Communication, *Erin M. Rowley, Kristen R. Moore, and Lauren Kuryloski, University at Buffalo (SUNY)*

Case Study #13: Environmental Engineering, *Jennifer Schneider and Lisa Greenwood, Rochester Institute of Technology*

Case Study #14: Computer Graphics Technology, *Rosemary Astheimer, Purdue University*

CASE STUDY #1

Mechanical Engineering

Erin Thomas, Iowa State University

SUMMARY

This case study will examine how standards information literacy instruction has been incorporated into the mechanical engineering design capstone course at Iowa State University. As the liaison librarian for mechanical engineering, introducing mechanical engineering students to library resources, including standards, and how to access and use them is a major component of my job. Standards information literacy instruction, in particular, was an important impetus for adding a one-shot library session to the capstone class, as there was very high demand for this content from both students and instructors. Over the course of several years, I developed and refined an instruction session for this course in which students learn strategies for finding standards, intellectual property, and scholarly resources to meet the needs of a wide range of engineering design projects. To develop this session, I built upon the work of Fosmire and Radcliffe [1] and the Iowa State University Library's existing standards LibGuide [2]. The session also has been shaped by extensive feedback from instructors and students.

TARGET AUDIENCE

The target audience is students in the mechanical engineering design capstone course. Students in this course are in their fourth and final undergraduate year at Iowa State University and are typically, but not always, mechanical engineering majors. One section of the course is multidisciplinary and is open to students from all engineering majors.

INFORMATION LITERACY LEARNING OUTCOMES

Learning outcomes for the standards information literacy session include that students will:

- be able to find standards online (including in ASTM Compass and ASME Codes and Standards for Academia) and in the library's print collection
- be aware that they can work with their librarian to identify appropriate standards for their project, and
- know that their librarian can add standards to the library's collection if they are not already available.

DESCRIPTION OF INSTRUCTION

Each semester, five to seven sections of the mechanical engineering capstone design course are offered, with each section consisting of about 40 students. These sections meet synchronously and in-person in a specialized engineering classroom. Students in the class are placed into groups and work with industry partners to design new products or update existing products. Class meetings for this course are nearly three hours long; the library session occurs during a portion of one class meeting for each section of the course and typically lasts between 45 and 60 minutes. Only part of this session can be focused on standards information literacy, as many other topics need to be addressed as well. Time for questions and answers is built into the end of each session.

For the library session, I visit each section in the engineering classroom to present basic information and demonstrations of relevant resources in real time. The standards part of the session begins with a brief discussion of what standards are and why they are important for engineering design students, practicing engineers, and consumers. After this, I introduce the students to a few places where they might encounter information about standards without specifically searching for standards, such as in their Compendex search results. I then provide some options for targeted searching for standards using subscription databases, typically ASTM Compass and ASME Codes and Standards for Academia. This includes a demonstration of how to locate and access the search tool as well as basic searching techniques.

I also demonstrate how to search for print copies of standards using our local discovery tool, Ex Libris Primo, and provide some basic information on how to find those documents in the library's Standards Center. Most importantly (and of perhaps greatest interest to the students, who are working with very limited project budgets), I wrap up the presentation by explaining that, as the standards librarian, I can purchase standards for the library's collection and will be happy to purchase any standards that they need to reference for their projects.

DESCRIPTION OF INSTRUCTIONAL ACTIVITIES/MATERIALS

The library session is supported by a course library guide, which takes the place of presentation slides and subsequently serves as the students' research hub throughout the semester. This guide includes information about and links to all the resources covered in the session, plus information on how to schedule a research consultation with the librarian. Additionally, this guide is automatically embedded into the course's LMS page via the SpringShare LTI, so it is easy for students to find any time they log into their course page.

ASSESSMENT

Assessment for the session is informal and is based on a combination of in-class observations, feedback from instructors, and conversations with students. Capstone students present posters and prototypes of their projects at an end-of-semester Design Expo, which is open to the public and provides an opportunity to chat directly with the students about their projects and their experiences finding and utilizing standards.

I have found that students are often very excited to have their librarian return at the end of the semester to see how things went. In general, they are more than willing to share their experiences if I demonstrate interest in their project outcomes and in improving the library session for future students. This has made the Design Expo both an excellent tool for obtaining student feedback on this particular session and for identifying opportunities to incorporate library instruction elsewhere in the mechanical engineering curriculum.

REFERENCES

1   M. Fosmire and D. F. Radcliffe, *Integrating information into the engineering design process.* West Lafayette, IN: Purdue University Press, 2014.

2   E. Thomas, "Library Guides: Standards & Specifications: A How-To Guide." https://instr.iastate.libguides.com/standards (accessed Jul. 13, 2021).

CASE STUDY #2

First-Year Engineering

Katie Harding, McMaster University

SUMMARY

This case study outlines an easy-to-implement active learning activity designed to introduce first-year engineering students to reading and understanding technical standards. Standards can be intimidating when students first encounter them, and this activity provides students with a low-stakes opportunity to examine a standard firsthand. This activity has been used in a first-year course titled "Integrated cornerstone design projects in engineering." The students in this course engaged virtually during one of their experiential lab sessions, but this activity also would be effective in an in-person environment.

TARGET AUDIENCE

The target audience for this activity is undergraduate engineering students who are encountering and needing to use technical standards for the first time, likely those in their first or second year of study.

INFORMATION LITERACY LEARNING OUTCOMES

This lesson has two learning outcomes. By the end of the session, students will be able to:

- articulate the purpose of a technical standard and its value to society; and
- locate needed information in a standard.

This lesson was developed for students who might need to consult technical standards in the process of completing a design project. I designed the lesson with the goal of helping students understand why the information in a technical standard is important and how they might use that information. I wanted students to feel comfortable engaging with technical standards and to feel confident in reading and finding information within.

DESCRIPTION OF INSTRUCTION

Before the activity begins, the instructor provides a short, 15–20-minute presentation that introduces students to technical standards.

In the activity, students view a freely available standard, the Ontario Drinking Water Quality Standards [1]. Working in small groups, students are provided with a list of questions about the standard to discuss and answer together. These questions prompt students to think about the standard's content, function, and value to society. They have 20 minutes to work together to answer the questions about the standard.

Once students have had a chance to consider and discuss these questions as a group, the instructor invites the groups to share their ideas in a discussion with the entire class or lab section. This discussion takes approximately 10 minutes.

DESCRIPTION OF INSTRUCTIONAL ACTIVITIES/MATERIALS

The instructional materials used in this lesson are:

- a presentation providing an overview of technical standards; and
- a handout that includes directions for accessing the technical standard that students will examine, and a list of questions for students to consider and answer in their group.

The presentation prepares students for the activity by providing a general introduction to technical standards. It includes information such as a definition of a standard and some examples, a description of the value of standards, an overview of the standards development process, examples of standards organizations, a discussion of some of the different terms students may encounter (standards, regulations, codes, guidelines, and specifications), and information about what standards students are able to access freely through their university library.

After the presentation, students form small groups and are given the handout, which includes a link to the Ontario Drinking Water Quality Standards, and the following questions:

1. How does this document protect the quality of Ontario's drinking water?
2. What kinds of substances are included in this standard?
3. Who would use this document?
4. Who benefits from the existence of this regulation? How do they benefit?
5. This document is both a standard and a regulation. What does that tell us about it?
6. What Act enforces this regulation? When was the Act created?

These instructional materials can easily be modified to use a different technical standard that will be of interest to students. The standard selected should be relatively short so that students can read through it in a brief amount of class time. It also should be easy for students to access, as many will not have any experience with accessing standards. Ideally, this means that the standard should be freely available online so that a link can be shared with students. Finally, the standard should relate to a topic that students are familiar with from their life experience, and from which it will be easy for students to understand the value of the standard and why it is important that it exists. The questions can be modified to be specific to whatever standard students are examining. They should be written in a way that guides students toward an understanding of the role of the standard in society, and how they might use it in their work as students and as engineers.

ASSESSMENT

After students have worked through the questions and answered them with their group, the instructor leads students in a discussion of their answers to the questions about the technical standard and responds to any student questions that arise. This provides an opportunity for some larger group discussion and ensures that students can hear perspectives from students in other groups and from their instructor.

Each group is required to submit their written responses to the questions. If they made a genuine attempt to answer the questions, they receive full marks for this activity. Because this activity will not negatively impact students' grades so long as they attempt to complete it, the focus is on exploring and learning about the standard, rather than on grading. Having each group submit their responses ensures that the instructor can check to see if they were able to correctly answer each question and if there are areas in which students would benefit from further instruction.

REFERENCE

1　*Ontario Drinking Water Quality Standards*, Ontario Regulation 169/03, 2002.

CASE STUDY #3

Law

Amanda McCormick, University at Buffalo

SUMMARY

This case study introduces students to how and why federal agencies use standards in regulations. Students will apply this learning in a homework assignment in which a law firm partner asks the student to research current regulations that may be applicable to a potential client. The assignment is designed for upper-level undergraduate students with an interest in law and policy.

TARGET AUDIENCE

Pre-law students in undergraduate political science courses, such as constitutional law, public policy, and legislative politics.

INFORMATION LITERACY LEARNING OUTCOMES

- Students will learn about standards.
- Students will learn about the federal rulemaking process, focusing on how federal agencies incorporate standards into regulations.
- Students will practice navigating a website that provides public access to regulatory materials (www.regulations.gov).
- Students will demonstrate an understanding of the use of standards in regulations.

DESCRIPTION OF INSTRUCTION

- Asynchronous: Prior to class, students will:
  - Read the U.S. Department of Transportation's overview of the rulemaking process (https://www.transportation.gov/regulations /rulemaking-process).
  - View the following videos: (1) What is incorporation by reference and why is it important? (https://www.youtube.com/watch? v=tFkXlqi_79U) and (2) Why do government agencies incorporate private standards? (https://www.youtube.com/watch?v=fk G4DHi5198).
- Synchronous:
  - Students are presented with an in-class lecture by the librarian on standards. The lecture covers what standards are, why standards are used (focusing on use by federal agencies), the standardization process, how to read a standard, and where to find standards, especially those available through library subscriptions. Approximately 45 minutes.

DESCRIPTION OF INSTRUCTIONAL ACTIVITIES/MATERIALS

- Lecture using PowerPoint presentation
- Videos
- Websites

ASSESSMENT

Homework assignment (also may be used as an in-class activity):

Directions: Please read the sample email from an attorney. Provide an email response to the partner, remembering to highlight the standards that you find in the rule(s)/proposed rule(s).

From: Senior Partner

Sent: September 7, 202- 7:51 AM

To: Associate <associate@worldwidelawfirm.com>

Subject: Research Needed ASAP

Associate,

The firm is hoping to land XYZ Automobile Co. as a client. From what I understand, XYZ is regulated by the National Highway Traffic Safety Administration. Please look at regulations. gov to research current rules and regulations (past 3 months or so) about automobiles. I need to see a few examples of the regulations to assess what is going on in the industry. Can you please research and email links to a few of the rules/proposed rules that you find relevant? I am particularly interested in regulations with standards that we may have to abide by.

Thanks,

Senior Partner

CASE STUDY #4

Health Sciences

Suzanne Fricke, Washington State University

SUMMARY

This narrative describes a case-based exercise for teaching fourth-year health science professional students about regulatory agencies and processes surrounding the development and safety of products that impact health or the provision of health care. Alternatively, this exercise could be expanded to interdisciplinary teams that include both health professionals and engineers.

TARGET AUDIENCE

Health science students or interdisciplinary teams of students who use or develop medical products, or products that impact health.

INFORMATION LITERACY LEARNING OUTCOMES

- In response to a given clinical adverse event, identify relevant laboratory testing and regulatory agencies, and find online reporting portals or registries.
- Evaluate currency, authority, and purpose of relevant online information from professional organizations, government agencies, laboratories, manufacturers, and scholarly publications.
- Identify potential standards or regulations relevant to development and safety of medical products, or products that impact the health of all species.

DESCRIPTION OF INSTRUCTION

This is a 100–120-minute synchronous exercise for small groups of 5–7 health science students in the final clinical year of their curriculum. It is case based with individual students selecting and presenting their decision-making process in response to pre-posted short clinical scenarios with regulatory aspects. These scenarios are provided with other course materials a week in advance of the flipped classroom exercise. Students take turns leading discussion of their cases for approximately 15 minutes per case. The exercise can be taught in person using a technology-enabled conference room with students in control of the keyboard, or online using screen share video conferencing. An experienced health information specialist with clinical knowledge, or working in collaboration with a clinician, fills a role as facilitator. Students are directed to state specific logistical actions taken in response to a food, drug, or device adverse event, and to demonstrate resources consulted, rather than focusing on broad or theoretical answers. This is taught within the context of a variety of other scenarios that include infectious disease reporting, travel regulations, and food safety, though it could focus exclusively on medical products. Scenarios arise from emerging public health concerns that appear in international, national, or local news, or cases presented to partnering diagnostic laboratories. Scenarios intentionally do not provide all details to encourage students to both adapt them to their own intended specialty or geographic area (country/state), and to encourage discussion of the role, if any, that circumstances play in decision making. The scenarios provided are short 2–3 sentence descriptions. One example is a dog presenting with hypoglycemia and liver failure secondary to ingestion of a drug formulation containing the sugar substitute xylitol, a food additive generally recognized as safe (GRAS) for human consumption by the FDA. Adverse events involving this product are optionally reported to the FDA through their consumer safety reporting portal. In response to consumer complaints, the recent Paws Off Act of 2021 (H.R. 5261) was introduced, calling for amendment of the Food, Drug, and Cosmetic Act to require labeling of xylitol products for their toxicity to dogs.

This session is guided using essential questions that frame the unit of study as a problem to be solved [1] These may include questions about diagnostic testing in response to an event, making this exercise applicable to both clinical and regulatory science learning. At the same time, it grants students decision-making opportunities and acknowledges them as future creators of systems [2]. In addition to guiding questions accompanying each scenario, overarching questions about what agency to report adverse events to or where to find information for consumers are presented at the beginning of the discussion to encourage exploration beyond the immediate case or client.

Students lead the group through online resources and data dashboards developed by professional organizations, laboratories, academic institutions, local health departments, and state and federal government agencies, as well as state legal code, scholarly publications, and consumer information. In the process, students evaluate information literacy concepts of currency, relevance, authority, and purpose. In using a flipped learning model, where students complete work in advance, this exercise attempts to provide students with low-stakes decision-making authority in an environment that provides immediate peer and mentor feedback [3].

While the session described is taught to health science students in a regulatory context, it could be expanded to interdisciplinary teams working to create or assure quality and safety of medical products. If teams have collective health science and engineering knowledge, potentially group presentations could look at both reporting systems for adverse events, and dive deeper into applicable standards for product development.

DESCRIPTION OF INSTRUCTIONAL ACTIVITIES MATERIALS

This exercise requires the development of brief case studies by a subject expert pathologist or clinician in collaboration with a librarian, or alternatively, a librarian with subject expertise. These cases change frequently in response to current events and changing regulations and reporting systems.

The session may benefit from the maintenance of a library guide of relevant government and organizational sites and relevant news stories. This site is not provided to students; however, it assists the facilitator in rapidly providing example resources to students in the midst of the educational session as is relevant to discussion.

ASSESSMENT

Student presentations and participation is assessed through observation, direct facilitator participation in discussion, and completion of a prescribed MedHub eValue web-based form. This evaluates communication, engagement, knowledge, and integrative abilities. Routine student evaluations of teaching indicate that participating health science students find the session "practical" and "clinically relevant," and value the open and informal facilitator and peer discussion.

REFERENCES

1   C. C. Kuhlthau, L. K. Maniotes, and A. K. Caspari, *Guided inquiry learning in the 21st century*, 2nd ed. Santa Barbara: CA; Denver, CO: Libraries Unlimited, 2015.

2   American Library Association, "Framework for information literacy for higher education," *Association of College & Research Libraries (ACRL)*, Feb. 9, 2015. https://www.ala.org/acrl/standards/ilframework

3   C. Duijn, L.S. Welink, M. Mandoki, et al., "Am I ready for it? Students' perceptions of meaningful feedback on entrustable professional activities," *Perspectives on Medical Education*. 6, pp. 256–264, 2017.

CASE STUDY #5

Business Management

Margaret Phillips, Heather Howard, Annette Bochenek,
and Zoeanna Mayhook, Purdue University

SUMMARY

This case is situated in the Purdue University Krannert School of Management undergraduate business program. Purdue is a Midwestern land-grant institution with an enrollment of 49,000+ students, of which approximately 2,700 are undergraduate business majors. For six years, business librarians have taught MGMT 110, a two-credit course titled "Introduction to Management and Information Strategies."

MGMT 110 is a required course for high-achieving first-year students directly admitted to the Krannert School of Management. The course broadly introduces students to the field of business while embedding information literacy competencies, such as finding, evaluating, and using information, within course content and assignments. Approximately 120 students are enrolled in MGMT 110 each year.

In 2021, the three business librarians who currently coteach MGMT 110 collaborated with an engineering librarian to create and deliver a standards education module in the course. The module was delivered during week 10 of a 16-week term, building on previous lessons that engaged the students with traditional academic (e.g., journal articles) and business (e.g., marketing reports) sources [1]. The authors were motivated by the results of a study by Phillips et al. [2], which found that even though standards are important for business practices, they are rarely integrated into undergraduate curricula at two top-ranked business schools. This module is derived from a lesson the engineering librarian (Phillips) cocreated for engineering technology students [3].

TARGET AUDIENCE

Undergraduate students in business management and related programs are the primary target audience. In addition to incorporating the module into a first-year business management course, the authors have used portions of the content in a cocurricular workshop for students in an undergraduate marketing association. Additionally, the authors believe the module can be easily modified to fit many other undergraduate programs.

INFORMATION LITERACY LEARNING OUTCOMES

The outcomes of this learning module are centered around introducing undergraduates in business to the topic of standards in the context of their discipline. The outcomes cover different levels of Bloom's Taxonomy domains, ranging from understanding to evaluating [4]. The four learning outcomes (LOs) are:

LO1: Describe basic information about standards (e.g., what they are, creators/authors, purpose).
LO2: Distinguish a standard from other information types (e.g., journal articles).
LO3: Assess how standards relate to consumer products.
LO4: Evaluate how standards apply to business decisions and operations.

DESCRIPTION OF INSTRUCTION

The instruction consisted of prework and an in-class active learning lesson:

The prework required the students to watch two introductory videos about standards (see "Description of Instructional Materials" below) and complete an assignment comparing and contrasting a standard document to a journal article on a similar topic [5]. The assignment asked students to consider questions, such as: What is the purpose of the two documents? Who wrote each document? Who is the intended audience? What information does each provide? How is each document structured? This assignment gave students an opportunity to read and engage with a standard before coming to class, considering it alongside an information source they were more familiar with for academic purposes. The assignment and an in-class debriefing align standards education with the "Authority is Constructed and Contextual" and "Information Creation as a Process" frames of the ACRL Framework for Information Literacy for Higher Education [6].

The 50-minute in-class lesson was delivered during the weekly lab (30 students/lab), where students were required to bring laptops to class. The authors created a lesson plan that began with an introduction by a business librarian, welcoming the students to the class and briefly discussing how standards relate to business [7]. Next, the engineering librarian reviewed the preclass assignment and delivered a brief presentation covering standards basics. Following the presentation, the engineering librarian introduced an activity where the students worked in groups (4–5 students, preassigned) to complete a worksheet considering how standards relate to a consumer product, such as a lawn mower or laptop. The engineering librarian and a business librarian then debriefed the activity, emphasizing how standards relate to the consumer products that the companies the students may work for after graduation manufacture and sell. Lastly, a business librarian introduced a group challenge homework assignment focused on standards (see "Assessment" below).

DESCRIPTION OF INSTRUCTIONAL MATERIALS

The authors used the following instructional materials:

- Video modules: as prework, the students are required to watch the introduction and modules 1 and 2 from the "Standards are Everywhere: An Information Literacy Approach to Standards Education" video series [8].

- PowerPoint slides for the in-class lesson [9].
- LibGuide that contains links to editable Google Docs activity sheets [10].

ASSESSMENT

The authors used formative and summative techniques to assess the LOs (see previous "Information Literacy Learning Outcomes"). Here are the assessment strategies used for each LO:

LO1: To gauge students' initial understanding of the topic, the authors created and assigned a premodule quiz that consists of 10 questions about the basics of standards [11]. The students complete this quiz in Qualtrics [12] before engaging in any of the instructional material. The same quiz was assigned to the students after the completion of the final assignment in this module (the group challenge) to gain insights into student learning. Also the authors used in-class discussion to identify learning gaps and address any misconceptions about the basics of standards.

LO2: The preclass assignment required students to compare and contrast a journal article and a standard. Additionally, the in-class discussion during the assignment debrief formatively assessed student understanding of this LO and was used to address any misconceptions.

LO3: During the in-class activity, the authors observed students engaged in group work with regard to this LO. Also the authors used in-class discussion during the activity debrief.

LO4: In the group challenge assignment [13], students were presented with a business case and related standards. Students were required to review the standards and make a recommendation regarding whether the business should pursue compliance.

REFERENCES

1   H. Howard, Z. Mayhook, and A. Bochenek, "MGMT 110 introduction to management and information strategies fall 2021 syllabus," *Library Faculty Staff Supplementary Materials*, Jan. 2021, [Online]. Available: https://docs.lib.purdue.edu/lib_fssup/9

2 M. Phillips, H. Howard, A. Vaaler, and D. E. Hubbard, "Mapping industry standards and integration opportunities in business management curricula," *Journal of Business & Finance Librarianship, 24(1–2)*, pp. 17–29, Apr. 2019, https://doi.org/10.1080/08963568.2019.1638662.

3 M. Phillips and P. McPherson, "Using everyday objects to engage students in standards education," in *2016 IEEE Frontiers in Education Conference (FIE)*, Oct. 2016, pp. 1–5. https://doi.org/10.1109/FIE.2016.7757698.

4 D. R. Krathwohl, "A revision of Bloom's Taxonomy: An overview," *Theory Into Practice, 41(4)*, pp. 212–218, Nov. 2002. https://doi.org/10.1207/s15430421tip4104_2.

5 M. Phillips, "MGMT 110: Comparing document types pre-class assignment," *Library Faculty Staff Supplementary Materials*, Jan. 2021, [Online]. Available: https://docs.lib.purdue.edu/lib_fssup/13

6 American Library Association, "Framework for information literacy for higher education," *Association of College & Research Libraries (ACRL)*, Feb. 9, 2015. https://www.ala.org/acrl/standards/ilframework (accessed Oct. 01, 2021).

7 M. Phillips, H. Howard, A. Bochenek, and Z. Mayhook, "MGMT 110: Standards module in-class lesson plan," *Library Faculty Staff Supplementary Materials*, Jan. 2021, [Online]. Available: https://docs.lib.purdue.edu/lib_fssup/10

8 M. Phillips, M. Fosmire, P. McPherson, A. Edmondson, and S. Gulati, "Standards are everywhere: An information literacy approach to standards education," Jun. 17, 2017. https://guides.lib.purdue.edu/NIST_standards (accessed Oct. 01, 2021).

9 M. Phillips, "MGMT 110: Standards module slide deck," *Library Faculty Staff Supplementary Materials*, Oct. 2021, [Online]. Available: https://docs.lib.purdue.edu/lib_fssup/14

10 M. Phillips and H. Howard, "Library guides: MGMT 110 standards module." https://guides.lib.purdue.edu/MGMT110 (accessed Oct. 25, 2021).

11 M. Phillips and H. Howard, "MGMT 110: Pre and post module standards quiz," *Library Faculty Staff Supplementary Materials*, Jan. 2021, [Online]. Available: https://docs.lib.purdue.edu/lib_fssup/11

12 "Qualtrics XM - Experience management software," *Qualtrics*. https://www.qualtrics.com/ (accessed Oct. 01, 2021).

13 H. Howard, A. Bochenek, Z. Mayhook, and M. Phillips, "MGMT 110: Standards group challenge assignment," *Library Faculty Staff Supplementary Materials*, Jan. 2021, [Online]. Available: https://docs.lib.purdue.edu/lib_fssup/12

CASE STUDY #6

Biomedical Engineering

Joanna Thielen and Jamie Niehof, University of Michigan

SUMMARY

The University of Michigan (UM) College of Engineering (COE) has 13 departments, 6 programs, and almost 11,000 students, served by 4 engineering librarians. The size and scope of this audience mean that in-person instruction for all students is not possible. Offering online learning objects is one way of providing information literacy instruction to more students in a way that is sustainable.

The engineering librarians recently worked with a library science graduate student to develop an "Introduction to Engineering Standards" module in our university's learning management system, Canvas. The module was implemented in three senior design courses (two biomedical engineering [BME] and one mechanical engineering [ME]) over several semesters.

TARGET AUDIENCE

The target audience for this module is upper-level undergraduate students and graduate students in the College of Engineering.

INFORMATION LITERACY LEARNING OUTCOMES

Our goal was to introduce the concept of standards to students in a way that would help them understand the importance and complexity of standards (illustrated through relevant and engaging examples) and the ability to find standards using UM Library resources. Our information literacy learning outcomes were:

1. Define what standards are and why they are used.
2. Identify and classify the different elements of a standard, such as scope, referenced documents, definitions, and conditions for use.
3. Locate the appropriate engineering standard using UM Library resources.

DESCRIPTION OF INSTRUCTION

The impetus for creating this module was anecdotal evidence that we were not teaching standards effectively during one-shot in-person instruction sessions due to time constraints and the complexity of standards as an information type. Owing to the large size and lack of research-based assignments in the COE curriculum before the senior level, few undergraduate students work with library resources in engineering courses until their senior design courses. When faculty mention standards in senior design courses, students are given little background information on what they are, why to use them, or how to locate them. To combat student information overload during library one-shot sessions, we decided to create the module as a self-paced, asynchronous, online learning object that could be directly integrated into courses or completed independently.

The completed module has now become part of assignments in three senior design courses. The first step in implementation was outreach. The engineering librarians emailed faculty they knew taught senior design courses or used standards in their courses, briefly introducing the new module (including a preview link) and summarizing how the module could easily be integrated into their course. The benefits of using the module are:

- It is an asynchronous, self-paced module, so students can go through it multiple times, at their time of need.
- Students have access to it until they graduate.
- Librarians' in-class time could be devoted to covering other information literacy topics.
- Librarians are able to import this module into Canvas course sites, including creating a new assignment or modifying an existing assignment, so implementing the module would be little work for the faculty.

In order to meet differing information literacy needs in multiple courses and departments, several further iterations of this module have been developed. Two BME faculty requested a version of the module that was BME-specific. The BME module has the same information literacy learning outcomes, but integrates more BME-focused standards examples (such as sterilization procedures for medical equipment) and includes several pages detailing BME standards that are commonly used in senior design projects. A shorter version of the module was developed for a sophomore ME design course. It removed the second information literacy learning outcome (finding standards through UM Library) because the faculty supplies students with necessary standards through Canvas. The shorter module provides scaffolding for when students take their senior design courses two years later.

The engineering librarians see the "Introduction to Engineering Standards" module as the beginning of a larger suite of engineering-specific modules. This is part of our broader instructional effort to shift instruction from in-person, one-shot sessions to asynchronous, online, and point-of-need. This allows us to reach more COE members and encourages them to learn at their own pace.

DESCRIPTION OF INSTRUCTIONAL ACTIVITIES/MATERIALS

Creation of the Canvas module took four months from brainstorming to completion, for a total of about 75–100 hours of work time. The four engineering librarians had the assistance of one School of Information graduate student, who worked on the module as part of her capstone project. The module consists of 21 pages, includes one quiz with three questions, and takes approximately 30–35 minutes for students to complete. Our goal was to make it general enough that any engineering department could deploy it, with real-world examples of standards such as elevator shaft specifications and shoe sizes. After the module was finished, we reviewed it for compliance using Web Content Accessibility Guidelines.

ASSESSMENT

During three semesters, 381 ME and 127 BME undergraduate students completed this module. In one semester, 47 BME undergraduate students completed the BME-specific version of the module. This means that over 550 engineering students have received a more comprehensive and consistent education on standards due to our modules. Exposure to these modules led faculty to implement other library modules in their design courses (such as "Academic Integrity" and "Introduction to Citation Management"). Anecdotally, all faculty members who integrated the modules into their courses were happy with them and planned to continue using them in future semesters, which we consider to be a prime indicator of success. One ME faculty member commented, "students handled the course expectations around standards better with the module than they had in past semesters." Furthermore, one BME faculty member described the most useful parts of the module are "how to access the standards databases from the library's website . . . and breaking down [the] different sections [of a standard]." The engineering librarians will be using the "Introduction to Standards" module as a starting point to create a suite of other online learning objectives and further develop our standards instruction.

ACKNOWLEDGMENT

This module is publicly available in Canvas Commons under the title "Introduction to Engineering Standards."

CASE STUDY #7

Civil Engineering

Xiaowei Wang and Yue Li, Case Western Reserve University

SUMMARY

This case study describes instructions for teaching undergraduate students how to search and identify appropriate standards, and how to effectively locate, read, understand, and implement them for the design of civil structures. Case-based lectures are given to guide the students through detailed interpretation of the design process of a real project example. A writing assignment on a different project example is designed as homework to enhance the acquisition and retention of knowledge for the students. Outcomes of the lesson are assessed through evaluation of the homework assignment.

TARGET AUDIENCE

This case is best suited for undergraduate level civil engineering students, specifically those in structural design (I and II) classes. It is recommended for classes of approximately 30-40 students.

INFORMATION LITERACY LEARNING OUTCOMES

- Students can demonstrate an understanding of the requirements of select standards
- Students can recall ways to get access to standards
- Students can demonstrate an ability to identify appropriate standards for their project
- Students are able to identify portions of a standard applicable to their project
- Students can demonstrate an ability to compose a technical report based on their reading and understanding of a standard.

DESCRIPTION OF INSTRUCTION

This class is a required course for undergraduate students in civil engineering. One aim of the course is to teach students how to identify and implement appropriate standards for the design of civil structures. Case-based teaching is applied through (1) lectures, and (2) writing assignments. Students are provided with a case study of a bridge design project with background information, as well as assignments to write a design process report that clarifies the key steps and associated information obtained from relevant standards. The writing assignment facilitates the acquisition of technical knowledge, enhanced reasoning and analytical skills, development of motivation, and awareness of non-technical issues, which are essential attributes of civil engineers.

DESCRIPTION OF INSTRUCTIONAL ACTIVITIES/MATERIALS

Lectures

BACKGROUND AND INTRODUCTION:

Based on a real project of structural design (e.g., a building or bridge), the instructor describes the background of the project and interprets the requirements from the client such as the state department of commerce or transportation.

IDENTIFICATION OF STANDARDS:

Students are guided to search and identify appropriate standards from accessible sources and assisted by the librarian if needed. Such standards should be able to provide critical information in terms of (1) design loads and their combinations, (2) construction material properties, and (3) calculation methods.

APPLICATION OF STANDARDS:

Taking critical structural components in the real project as examples (e.g., columns in buildings or bridges), students are guided to locate specific portions of the above-identified standards and to understand how to read and apply them for the design of the components.

Assignment: Design Process Report

WHO YOU ARE:

A structural engineer at a consultant firm located in southeast coastal regions of the United States, e.g., Charleston, South Carolina (SC), which is prone to multiple natural hazards such as hurricanes, storm surges, and earthquakes.

THE SITUATION:

Your supervisor assigns you to a project as a high priority, the South Carolina Department of Transportation (SCDOT), for the design of a reinforced concrete bridge along the coast region of Charleston, SC. The SCDOT has provided their requirements on the bridge in terms of its role in the local and state transportation networks, traffic flow capacity, expected service life, etc. They also provided some recommended design codes and standards, but your supervisor reminds you that the codes and standards recommended alone may not be sufficient to complete the design.

OTHER DETAILS:

Based on your past work experiences, you understand that you will need to use ASCE 7-16 (2016), Minimum Design Loads for Buildings and Other Structures; ACI PRC-343-95 (2004), Analysis & Design of Reinforced Concrete Bridge Structures; and AASHTO LRFD Bridge Design Specifications (2011).

To determine the design loads and their combination of the reinforced concrete bridge under the potential of a multi-hazard scenario, ASCE 7-16 Standard (2016) is used. As for the design of reinforced concrete, ACI PRC-343-95 (2004) is adopted to determine the design strength of rebars and concrete. Accordingly, AASHTO LRFD Design Specifications (2011) are complied with to design the bridge.

DELIVERY:

Write a technical report that describes the key steps and the associated requirements and criteria specified in associated standards. In line with the lecture and class discussions, the technical report needs to contain the following contents:

PROJECT DESCRIPTION:

Describe the project and highlight the design requirements

STANDARDS IDENTIFICATION:

List the required standards and justify their necessities.

STANDARDS IMPLEMENTATION:

Interpret the key steps in the design process. In particular, the report needs to demonstrate the recommended bridge type(s) from the applied standards, together with the evaluation of design requirements.

ASSESSMENT

Assignment: Design Process Report

To demonstrate the abilities students learned from this course, a written home-work of a technical report on the design process of the given project example is assigned. Students are expected to meet the following assessment criteria.

Assessment Criteria:

- Understand the design requirements of their project
- Develop a list of relevant standards
- Locate the relevant sections in the standards
- Understand the identified sections of standards and their significance to the key steps in the design process
- Final report, 500-750 words with an introduction, standards identification, standards implementation sections

CASE STUDY #8

Electrical Engineering

Seyed Hossein Miri Lavasani, Case Western Reserve University

SUMMARY

The purpose of this case study is to introduce the basics of wireless communication standards to graduate-level electrical engineering students.

TARGET AUDIENCE

This case study is best for graduate-level electrical engineering students, ideally those in a design course with roughly 10–15 students.

INFORMATION LITERACY LEARNING OUTCOMES

Following the case study, students should be able to:

- understand the basics of wireless communications and become familiar with different standards (for short- and long-range applications);
- understand the basics of the IEEE 802.11 standard, which is used for Wi-Fi and wireless local area networks (WLAN); and
- use the standards and obtain key information needed to design specific circuit blocks used in WLAN-compatible wireless transceivers. This included knowing how to access the latest release of the standard from the IEEE standards website (IEEE Xplore) and identifying the portion of the standard needed for their project.

DESCRIPTION OF INSTRUCTION

Students are presented with standards information embedded within several lectures throughout the course. The related topics covered in the lectures include:

- What are WLAN systems and what IEEE standards regulate their usage?
- What are the key metrics in the standards and how are they used?
- How to search for standards on IEEE Xplore and distinguish between older and newer releases.
- How to read through the standard and find the key information needed for the design of the appropriate circuit block.

Students are encouraged to use this standard for their course project as well as to participate in a course assignment to contribute to the next release of the WLAN standard.

SPECIFIC ACTIVITIES/MATERIALS

Writing Challenge: Standards Case Study

WHO YOU ARE:

A mid-career R&D engineer in a large semiconductor company, who is recently assigned to the internal standards development group, Intel Inc.

THE SITUATION:

The group leader will ask you to participate in the development of the next release of the IEEE WLAN standard based on the existing (802.11ax). There is a strong demand for a higher data rate and the available sub-6GHz electromagnetic spectrum is very limited. While your experience in standards development is very limited, the lead really values your insight due to your extensive R&D experience on high-speed wireless transceivers throughout your career.

THE PRESENTATION TO THE APPROPRIATE IEEE STANDARDS COMMITTEE:

The team will present system-level analysis along with characterization data obtained from the evaluation of an engineering sample unit designed to comply with the proposed new WLAN standard. Since the committee has likely gone through several prior releases of the proposed standard, they may have detailed feedback along with recommendations and testing results from other company members to share and discuss. They may also provide a list of recommended features based on their interactions with original equipment manufacturers (OEMs) such as Apple. The outcome of the meeting will be documented and divided into technical features needed to be implemented in each section such as the physical (PHY) layer, medium access control (MAC) layer, and so forth.

TESTING DETAILS:

Per the Federal Communications Commission (FCC) requirements, there is strict in-band and out-of-band electromagnetic emissions requirements for any WLAN-enabled device across the United States. In addition to that, there are other requirements such as blocker compliance, coexistence, and electromagnetic safety limits that need to be considered while making a recommendation to the standards committee. Therefore, a creative solution involving changes to various parameters such as the carrier frequency, bandwidth, modulation type, output power level, and output spectrum, and so forth may be necessary.

YOUR CHALLENGE:

Write a technical report that contains the following sections:

- A recommendation of a set of appropriate modifications to the existing standard to support data rates in excess of 10 Gb/s while complying with FCC requirements. The reasoning and evaluation plans should be clearly stated in the document.

ASSESSMENT

Writing Assignment: Standards Recommendation Report

Proper documentation of activities, including progress reports, design proce-dures/reviews, and the evaluation results play a key role in your success as an en-gineer in the industry. This assignment asks that you locate, read, and research the existing IEEE 802.11 standard to extract the key information needed for the design of the particular radio frequency (RF) circuit block that was assigned to you in the course project. It also requires you to develop recommendations to improve the existing WLAN standard, allowing WLAN wireless transceivers to achieve data rate beyond 10Gb/s.

REQUIREMENTS:

- Location and analysis of the WLAN standards (IEEE Xplore standards database).
- Research and extraction of key information needed to determine the specifications for the project.
- Drafting of relevant recommendations in memo format to achieve 10Gb/s.
- Preparing a document that clearly articulates the changes to the existing release of the standard as well as new evaluations/tests needed for charac-terization of new WLAN wireless transceivers.
- Developing a testing/evaluation plan that allows for effective evaluation and compliance verification of the wireless transceiver with respect to the FCC requirements.
- Writing a final report, 500–750 words with appendices, appropriate cita-tions, and these sections:
  - Introduction
  - Overview of Relevant WLAN Standard Releases
  - Recommended Modifications
  - Required Evaluation and Testing Procedures

CASE STUDY #9

Fire Science

Ya-Ting Liao and Daniela Solomon, Case Western Reserve University

SUMMARY

A university professor and a campus engineering librarian joined forces to promote the training of engineering standards in an academic curriculum. The objectives are to increase awareness of engineering standards and to emphasize the importance of standards on project development. They designed a training module that consists of a one-time lecture and a course project. The lecture presents a general introduction on standards, informs students of the characteristics of U.S. standardization, teaches strategies for finding and using standards, and highlights standard resources available at the campus library. The lecture is followed by a project assignment. In the project, students select a product related to the class topic and identify at least three standards applicable to that product. Students are required to investigate and present to the class how these standards have impacted the design of the product. This case study documents the content of the training module and how it was executed in a graduate-level course.

INFORMATION LITERACY LEARNING OUTCOMES

- Students know how to identify and apply relevant standards for their products.
- Students are familiar with the campus resources to access the standards.
- Students understand the value of standards for engineering design and applications.
- Students are able to explain the impact of selected standards on product development.

TARGET AUDIENCE

Mechanical and aerospace engineering and macromolecular/polymer science fire dynamics classes.

BACKGROUND/CONTEXT (CLASS SIZE, PROGRAM, ETC.)

A three-credit-hour graduate-level course, dual-listed in the Department of Mechanical and Aerospace Engineering and the Department of Macromolecular Science and Engineering. The course is offered in the fall semester every year and meets twice a week for 75 minutes each time. Typical student enrollment is approximately 10 students, including graduate students and upper-class undergraduate students from various departments in the Case Engineering School.

DESCRIPTION OF LESSON

The course syllabus is updated to include a one-lecture (75-minute) standards training session and a team project. The engineering librarian gives the training session. The lecture explains what standards are, why standards are used, the standardization process, how to read a standard, and how to search for standards with emphasis on campus resources (e.g., library subscriptions). A project assignment follows the training session. Students work in teams (2–3 students in a team). Each team selects a product related to the class topic, identifies at least three standards applicable to that product, and studies how these standards have impacted the product's design. The lecture and the project assignment occur near the mid-term of the semester. The timing of the project assignment is to ensure that the students have adequate background in the course topic (in this case, fire dynamics) and can correlate the fundamentals and practical application. The timing also allows students to have enough time to work on the project before the semester ends. In the last lecture of the semester, each team summarizes its project results, delivers a 20-minute presentation to the class, and submits a final report (maximum 5 pages).

SPECIFIC ACTIVITIES/MATERIALS

- Lecture using PowerPoint slides
- In-class use of live polls (e.g., Poll Everywhere) to determine the understanding of critical content
- Project assignment

Students work in groups to select a product and identify three applicable standards. Each group will summarize the project results in a 20-minute presentation and a final written report (maximum 5 pages). The presentation should include the following:

1. Description of the product.
2. A list of three standards applicable to this product.
3. Details for each standard on the list (e.g., SDO, type, applicability, specific requirements).
4. Discussions on how each standard has impacted the product development.
5. Reflections on the use of selected standards.

ASSESSMENT

Assignment: Standards In-class Presentation

The effectiveness of the standard training is mainly evaluated by the project outcome. During the student presentation and in their final written report, the following items are examined.

- Whether or not the students identify appropriate, relevant, and updated standards for the product they present.
- Whether or not the students successfully acquire the identified standards.
- Whether or not the students analyze the normative requirements of each standard and their impact on the product development.

In addition, the effectiveness of the training module is evaluated through students' feedback in the anonymous teaching evaluation conducted by the university. The following questions are added to the standard evaluation.

- Do you agree that the guest lecture on the technical standard was effective and increased your understanding of standards?
- Do you agree that the project on the technical standards helped you to learn how to identify and apply standards?
- Please feel free to share any feedback or suggestions on the technical standard training.

For the first two questions, students are asked to choose one of the following: SD: Strongly Disagree; D: Disagree; M: Mixed Feelings; A: Agree; SA: Strongly Agree; N: I choose not to respond. For the third question, students can provide any comments in a text box.

CASE STUDY #10

Transportation Engineering

Thomas Abdallah, Metropolitan Transportation Authority Construction & Development, New York City Transit; Yekaterina Aglitsky, Metropolitan Transportation Authority Construction & Development, New York City Transit; Shirley Chen, New York City Mayor's Office of Environmental Remediation; Maria Cogliando, New York University; Louiza Molohides, Columbia University; and Angelo Lampousis, Department of Earth and Atmospheric Sciences, City College of New York, City University of New York

SUMMARY

This case study is suitable for undergraduate courses consisting of 25–40 students majoring in geosciences and/or engineering. Depending on the course focus, this case study may serve as a standalone lecture or a lecture series.

INFORMATION LITERACY LEARNING OUTCOMES

- Develop an ability to identify standards applicable to the evolving needs of sustainable transportation appropriate for densely populated urban areas.
- Develop an ability to apply standards-based engineering design solutions in legacy transportation infrastructure that balance public health, environmental, and economic constraints.
- Develop an ability to collaborate as a team on the adoption and implementation of an environmental management system as prescribed by international standards in order to meet environmental objectives.
- Develop an ability to adopt and apply management tools as informed by mainstream international standards to accomplish a process cycle of plan, do, check, and act.

TARGET AUDIENCE

This case study is aimed at those in the discipline(s) of transportation engineering, environmental management, and/or sustainability. It is suitable for classes in geosciences and/or engineering.

DESCRIPTION OF LESSON

An opening lecture will focus on the large-scale adaptation of international environmental management standards. A case in point is the Metropolitan Transportation Authority (MTA) of New York City Transit (NYCT), the largest transit system in the United States and the first public transportation company in North America certified to the ISO 14001 Environmental Management Systems.

Since 1999, MTA NYCT's Capital Program Management (CPM) has been certified under the ISO 14001 Environmental Management Systems (EMS) standard. In 2020, CPM underwent a major organizational transformation and is included in the new structure of Construction and Development that is responsible for projects for a wider range of MTA agencies. As of 2021, the scope of ISO certification continues to encompass planning, project development, design, and construction management of capital projects on the NYCT service territory. The certification is projected to extend to other MTA agencies, including MTA

Long Island Railroad, Metro-North Railroad and Bridges, and Tunnels over the period 2021–2024.

The lecture covers the components of an environmental management system (EMS) that includes management tools, knowledge of environmental policy, aspects and impacts, monitoring and measuring (e.g., heating fuel usage, electrical consumption, saving water, greenhouse gas releases), and the plan-do-check-act sequence as per ISO 14001:2015.

The American National Standards Institute (ANSI) is the U.S. member body of the International Organization for Standardization (ISO). Upon request, ANSI has been authorized to provide complimentary access for students and faculty to selected standards currently available in the ISO collection, including ISO 14001:2015.

SPECIFIC ACTIVITIES

At the conclusion of the opening lecture, students are challenged through a series of images and animations to identify the application of the ISO 14001 EMS standard by MTA NYCT at different levels, ranging from baseline compliance to sustainability initiatives and mitigation measures. Breakout rooms consisting of 2–4 students will then focus on different components of MTA NYCT infrastructure. These include station and terminal environments, support structures such as substations, ventilation facilities, pumping facilities, transit infrastructure lines, and the necessary trackbed and rail system to propel trains that carry passengers from point to point within cities. Student groups may also focus on ancillary facilities such as train storage yards, maintenance shops, and bus depots. At the conclusion of the class, student groups report their findings to the whole class and relate any anticipated benefits of particular sustainability initiatives or mitigation measures that they were able to identify.

ASSESSMENT

A writing assignment will challenge students to explore the potential application of the ISO 14001 EMS standard in other contexts. Students are free to select urban environments from anywhere in the United States and internationally, as long as a preliminary literature review indicates that the ISO 14001 EMS standard has not been adopted to date. Students then make calculations and projections similar to the ones demonstrated through the opening lecture for MTA NYCT for their focus location. The deliverables include maps and other site-specific data, as well as volume projections of any greenhouse gas reductions or other environmental benefits through the potential adoption of an EMS. The scope of this writing assignment can be adjusted based on the scientific background of particular student groups and on the desired duration of this module.

ACKNOWLEDGMENT

This educational module was made possible by the National Institute of Standards and Technology (NIST) United States Department of Commerce (DoC) Standards Services Curricula Development (SSCD) Cooperative Agreement Program. NIST Award Number 70NANB16H266 for the period 11/01/2016—10/31/2018. Dr. Angelo Lampousis, Principal Investigator.

FURTHER READING

T. Abdallah, "Chapter 9—Environmental management systems," in *Sustainable Mass Transit*, T. Abdallah, ed. Elsevier, 2017, pp. 123–139. https://doi.org/10.1016/B978-0-12 -811299-1.00009-5.

A. Lampousis, "On the pursuit of relevance in standards-based curriculum development: The CCNY approach," *Standards Engineering: The Journal of the Society for Standards Professionals, 69(4),* July/August 2017, pp. 1–6.

Environmental management systems—Requirements with guidance for use, ISO 14001:2015(en), International Organization for Standardization (ISO), Geneva, Switzerland, reaffirmed 2021.

CASE STUDY #11

Mechanical and Aerospace Engineering

Daniela Solomon and Ya-Ting Liao, Case Western Reserve University

TARGET AUDIENCE

This case is intended for mechanical and aerospace engineering heat transfer classes. The curriculum of this course covers fundamental principles of heat transfer by conduction, convection, and radiation, and applications of these principles to the solution of engineering problems. It is a required course for mechanical engineering and aerospace engineering. The class typically consists of approximately 80–100 upper-class undergraduate students. While most students are from the Department of Mechanical and Aerospace Engineering, there are students from various departments in the School of Engineering. The class meets three times a week for 50 minutes each time.

INFORMATION LITERACY LEARNING OUTCOMES

- Students can discuss the value of standards for engineering design and applications.
- Students can recall ways to get access to standards.
- Students demonstrate an understanding of how content of a standard is organized and the role of each section.
- Students demonstrate an understanding of how to interpret a standard using language clues included in the standard.

DESCRIPTION OF INSTRUCTION

The course syllabus includes a one-lecture (50-minute) session dedicated to standards training through lecture and an in-class exercise in interpreting the requirements of two related ASTM standards. The lecture explains what standards are, why standards are used, the standardization process, how to read a standard, and how to search for standards with emphasis on campus resources (e.g., library subscriptions). Each section in the lecture is followed by an online poll that asks a question on the topic of the section to assess students' understanding (formative assessment techniques). The lecture is followed by an in-class or take-home exercise that consolidates some of the notions introduced in the lecture.

At the end of the lecture, students are prompted to use the ASTM Compass database available on the library website and find two standards, E2585—Standard Practice for Thermal Diffusivity by the Flash Method, and E1461—Test Method for Thermal Diffusivity by the Flash Method. Once they have opened the full text of the two standards, students are invited to answer several questions that bring their attention to specific elements from the content of the two standards. Through the questions asked, students self-navigate through all elements of the standards and reflect on their interpretation. They also have the opportunity to ask questions and clarifications. When the class moved to an online format during the COVID-19 pandemic, the in-class exercise was adapted into an online quiz delivered as a homework assignment.

DESCRIPTION OF INSTRUCTIONAL ACTIVITIES/MATERIALS

- Lecture using PowerPoint
- In-class use of live polls to determine the understanding of critical content
- Online access to standards ASTM E2585—Standard Practice for Thermal Diffusivity by the Flash Method, and E1461—Test Method for Thermal Diffusivity by the Flash Method
- In-class (or take-home) exercise in interpreting two ASTM standards related to heat transfer

FORMATIVE ASSESSMENT QUESTIONS

1. Which of the following is NOT true about the benefits of standards?
   a. Specify requirements for operation, quality, safety
   b. Reduce product development costs
   c. Increase transaction costs
   d. Create a level playing field for producers
   e. Create common language
2. What does voluntary standard mean? Choose all that apply.
   a. Organizations volunteer to participate in the development process
   b. Developed by a recognized body
   c. Compliance is voluntary
   d. Market-driven
   e. All of the above
   f. None of the above
3. Which of the following statements are true? Choose all that apply.
   a. Compliance is determined based on self-testing
   b. Compliance is determined by an independent certification body
   c. Conformity is determined based on self-testing
   d. Conformity is determined by an independent certification body

POST-LECTURE EXERCISE QUESTIONS

1. What SDO has developed these standards?
2. In what year was the E1461 standard published?

3. Are these standards different from each other? In what way?

4. What are the limitations in applicability of the practice presented in the scope of the E 2585 standard?

5. Are there any imposed design limitations for the testing apparatus included in the E 1461 standard?

6. What does the use of "may, need not" represent in the context of any standard? Find a couple of examples in the E 2585 standard and analyze the differences compared to "shall, shall not."

7. Is compliance to both standards required simultaneously? How can you determine?

8. What units of measurement are mentioned in these standards?

9. What is the repeatability value estimated for the testing method presented in E 2585 standard?

10. What ASTM committee developed these standards?

CASE STUDY #12

STEM Communication/ Technical Communication

Erin M. Rowley, Kristen R. Moore, and Lauren Kuryloski, University at Buffalo (SUNY)

SUMMARY

Reading and writing technical communication are skills used by engineers constantly in their careers. Technical standards are an example of technical communication that many engineers use frequently in their careers. The purpose of this case study is to introduce undergraduate-level engineering students to standards and provide guidance on how to read this special form of technical communication [1]. Further, it addresses the skill of writing technical communication, as students must disseminate information from the standard to write a recommendation report.

TARGET AUDIENCE

Engineering undergraduate-level students in any discipline, typically those in their second or third year; STEM communication/technical communication course. This case study was designed for smaller classes of between 10 and 40 students.

INFORMATION LITERACY LEARNING OUTCOMES

- Students can discuss the value of standards for engineering design and applications.
- Students demonstrate an understanding of the requirements of select standards.
- Students can recall ways to get access to standards.
- Students are able to identify portions of a standard applicable to their project.
- Students can demonstrate an ability to compose a technical recommendation report based on their reading and understanding of a standard.

DESCRIPTION OF LESSON

Students are presented with an in-class lecture by the engineering librarian on standards. The lecture covers what standards are, how they are used, how to search for standards (especially those available through the library's subscriptions), and how to read through a standard.

Students are provided with a case study to provide background information, as well as a course assignment to write a standards recommendation report.

SPECIFIC ACTIVITIES/MATERIALS

The case study and assignment below are distributed to students at the beginning of the unit.

Writing Challenge: Standards Case Study

WHO YOU ARE:

An entry-level product test engineer at a mid-sized consumer products testing laboratory, ProAssess.

THE SITUATION:

Your supervisor assigns you to a special case for a high-priority client, a toy retailer called E&E Toys. The client has designed a new toy and has come to your company to have it tested, as the law requires, to ensure it meets applicable toy safety requirements. While you have only worked at ProAssess for about 18 months, your supervisor says she trusts your judgment to review the toy in question based on your stellar track record.

THE CLIENT AND THE PRODUCT:

The toy has gone through several phases of design, but they do not quite have a functioning sample yet; however, they have detailed drawings and pictures of the prototype to share. They also have a list of the toy features, but your supervisor warns you that the feature list is from the client's marketing department, and therefore may not be a complete list of technical functions.

You have worked with this client before, but this is the first new product you have reviewed. However, based on your past experience with the client, you know that E&E Toys does not have a large product testing budget, and prefers to test for only those requirements that are truly necessary.

OTHER DETAILS:

ProAssess purchases industry standards and test methods from various standards organizations in order to conduct business. In addition, your company subscribes to the ASTM Compass database, which provides electronic access to all current and past versions of ASTM standards, which are extremely important to the consumer products testing industry.

Based on your past work experiences, you know that you will need to consult ASTM F963-17, Standard Consumer Safety Specification for Toy Safety; however, you no longer have a copy, so you will have to locate this current version of the standard using the ASTM Compass database.

TESTING DETAILS:

Per the U.S. Consumer Product Safety Improvement Act of 2008 (CPSIA), all sections of ASTM F963 are mandatory; however, not all toys possess all the various features and functions that are detailed in the test standard.

While you have not been given a specific budget to work with for this assignment, as far as what the client is willing to pay, your supervisor reminds you that each test method listed in ASTM F963 would be a separate line item on the bill to E&E Toys. She understands the client does not like to spend money on product testing but wants to ensure that ProAssess is not held liable for failing to recommend potentially appropriate tests.

YOUR CHALLENGE:

Write a technical report that contains the following sections: A recommendation of tests and other requirements from the applicable toy test standard, ASTM F963, as well as your evaluation of the specific toy. Remember to provide reasoning as to why you are recommending specific tests or markings.

ASSESSMENT

Students are assessed on their knowledge of reading and understanding technical communication (the standard) through a writing assignment in the form of a recommendation report.

Writing Assignment: Standards Recommendation Report

As an engineer, particular types of writing and communication will infiltrate your day-to-day life, including daily communication practices (like email or meeting briefs), understanding and responding to technical standards, and developing recommendations based upon both. This assignment asks you to read, research, and develop recommendations about the design of a particular product based upon the technical documents and research you have done.

REQUIREMENTS:

- Location and analysis of standards (using ASTM Compass database)
- Development of relevant standard section list
- Drafting of recommendations in memo format

- Drafting of introduction, recommendations, and other sections
- Final report, 500–750 words with appendixes, appropriate citations, and these sections:
  - Introduction
  - Product description and analysis
  - Overview of relevant standards
  - Recommended tests
  - Required tests

REFERENCE

1   E. M. Rowley, L. Kuryloski, and K. R. Moore, "Extending the role of the library and librarian: Integrating alternative information literacy into the engineering curriculum," presented at the 2020 ASEE Virtual Annual Conference Content Access, Virtual online, June 22, 2020. [Online]. Available: https://peer.asee.org/34656

CASE STUDY #13

Environmental Engineering

Jennifer Schneider and Lisa Greenwood, Rochester Institute of Technology

TITLE OF CASE STUDY

Standards-Based Curriculum Across Risk Management System Domains: Health, Safety, Environmental, and Community-Sustainable Development Standards

SUMMARY

Moving beyond compliance with laws and regulations to continual improvement in our facilities and communities relies upon sustainable, earth- and person-centered strategies, goals, and operations. Implementing consensus standards, such as risk-based management systems standards created and led by the International Organization for Standardization (ISO) is especially relevant for the environmental, health, and safety (EHS) management profession. Three of these in particular, ISO 14001, 37101, and 45001 [1, 2, 3], shape how our society conducts itself, and how EHS professionals create value for society.

Three corresponding standard-specific modules were created for integration across undergraduate (UG) and graduate (G) curricula in the Rochester Institute of Technology (RIT) BS in environmental sustainability health and safety and MS in environmental health and safety management, and shared with other programs and instructors [4, 5]. Each module was structured with an overview, framework and benefits, educational content outline, and resources sections; and each was validated by professors across broad domains. Intentionally, the modules were built to be customizable across a range of domains and ability levels [6, 7]. The educational goal is to cultivate professional capacity to apply these standards with a real-world, applied focus, with the tools, techniques, and capabilities to be internal change agents for the greater good.

TARGET AUDIENCE

The modules were created in a flexible, customizable format that allows a professor to extract and implement course-relevant content across technical and business domains, as student learning outcomes require. The learning outcomes and content were labeled by student capacity, whether UG- or G-focused, or both. Graduate-targeted curriculum requires students to engage in the application through analysis and evaluation, while undergraduate and shared (UG and G) curricula focus on the identification and explanation of the key concepts within the standards, with application in specific contexts.

INFORMATION LITERACY LEARNING OUTCOMES

The curricular content was designed to build student capacity to apply risk-based EHS management standards that enable organizations to address gaps in regulation from a global perspective and provide a means for consistency in operations, as well as a means of managing risk. The generalized learning outcomes across the three-management system standard modules are:

1. Identify and describe the framework for the standard (UG and G).
2. Explain (UG and G) and apply (G) the key requirements associated with management system standard elements.
3. Analyze (UG and G) and evaluate (G) issues and strategies for specific cases and applications of the standards.

DESCRIPTION OF INSTRUCTION

Teaching and learning of the curriculum is framed in a Microsoft PowerPoint format, with supplemental materials that include various links to additional resources for the professor and student to enhance understanding. While the content can be reviewed asynchronously, it is the intent of the authors to teach this in a supported format. The modules also provide activities, assignments, and assessment tools to advance learning and integrate assessment opportunities. Since the modules and content are customizable, teaching time will vary, as short as a one-hour overview of each standard to as much as the majority of a semester course focused on standards.

TABLE 13.1 *Case Study Instruction Components and Descriptions*

Component	Description
Module Roadmap	Executive summary with introduction and overview of the module; module learning outcomes, description, and rationale
Standards Overview	Summary and scope of the standard the module addresses
Educational Content	PPT lecture slides with guided activities, exercises, and lecture notes; supplementary resources, e.g., readings, links to materials/tools; answers to frequently asked questions from the student perspective; example discussion questions, assignments, and exercises; website posting on RIT Collaboratory for Resiliency & Recovery [8]
Module Assessment	Assessment tools and methods to measure module effectiveness

ASSESSMENT

Class exercises and assignment options are incorporated into the curriculum to promote learning and competency building, including content-driven questions and discussion prompts, simulations, small group work, and comprehensive case-based projects. Particularly for these risk-based standards, students must recognize that performance is process-driven; therefore, ultimate success is driven by a strategic performance approach, informed by a set of requirements. Similarly, evaluation of student learning growth is guided by assessment rubrics informed by performance [4, 5]. As an example, in graduate-level small group simulations, students implement specific requirements of the ISO 14001 and 45001 standards [1, 3, 8], determining a strategy for a fictitious company. After the class presentation, an instructor-led debriefing session compares results, critiques group decisions, and assesses the extent to which student artifacts met the requirements of the standards. Students are then given the opportunity for revision before final submission based upon their rubric score, peer critique, and instructor feedback [4–7].

Student Engagement Example — Implementing EHS Risk Management Standards

In RIT's introductory graduate-level course, EHS management, class activities and assignments build upon each other to develop student capacity related to EHS risk management. Preliminary assignment effort is reinforced through engagement with the instructor and peers designed to elicit and reinforce learning [9], as outlined in Table 13.2.

ACKNOWLEDGMENT

This material was prepared by RIT under award 70NANB16H268 from the National Institute of Standards and Technology, U.S. Department of Commerce. Statements, findings, conclusions, and recommendations are those of the authors and do not necessarily reflect the views of the National Institute of Standards and Technology or the U.S. Department of Commerce.

REFERENCES

1 *ISO 14001:2015 Environmental management systems—Requirements with guidance for use*, International Standard ISO, Geneva, Switzerland, 2015.

2 *ISO 37101:2016 Sustainable development in communities—Management system for sustainable development—Requirements with guidance for use*, International Standard ISO, Geneva, 2016.

3 *ISO 45001: 2018 Occupational health and safety management systems—Requirements with guidance for use*, International Standard ISO, Geneva, 2018.

4 L. Greenwood, J. Schneider, and M. Valentine, "Setting a course for student success: Standards-based curriculum and capacity-building across risk prevention

TABLE 13.2: *Student Engagement Example—Activities and Descriptions*

Curricular Activity	Description
Preliminary Assignment	Students access and review ISO 14001 and 45001 standards through the ANSI University Outreach Program and read assigned supplementary literature.
Lecture	Instructor guides students through risk identification, assessment, and control requirements of the standards, introducing related approaches and tools.
In-class Exercise: Turn to Partner	Students examine images of situations in the workplace, work in pairs to identify EHS hazards and risks, and share with the class. Instructor assists with prompts as needed to help students consider each activity and engages students in discussion about potential impacts [9].
Individual Assignment	Students respond to questions designed to build their standards vocabulary, compare the risk management requirements in multiple standards, and engage with the guidance and supporting materials to enhance their understanding of the intent of the standards, their similarities, and key differences.
Online Discussion	Students select one concept or requirement from the standards and discuss why it is important for effective EHS management, responding to prompts from the instructor and questions from peers.
Project Part I	Students work in groups outside of class for a multipart project focused on implementing business risk management standards. In Part I, they research the business to gain an understanding of its context and how its activities can affect workers and the environment. Based on this information and their understanding of the standards, students identify the organization's EHS risks and develop criteria for assessing them. The instructor provides feedback via a detailed assessment rubric [4–7].
Project Part II: Workshop	Students participate in a one-day summative workshop. Building upon the previous assignment, students apply their criteria to assess the organization's EHS risks and determine those that need to be addressed, develop a strategy to address these risks as required by the standards, and present their output artifacts to their peers [4–7].
Debrief and Critique	At the end of the workshop, guided by the project assignment rubric, students discuss and critique the results, and receive constructive feedback from peers [6, 7].
Final Assessment	Using the project assessment rubric, the instructor assesses student mastery of standards content based on their final artifacts.

management system domains," in *2018 ASEE Mid-Atlantic Section Spring Conference*, 2018.

5   L. Greenwood, J. Schneider, and M. Valentine, "Environmental management systems and standards-based education: A modular approach," in *Charting the Future: Environment, Energy & Health, A&WMA 111th Annual Conference & Exhibition*, Hartford, CT, 2018.

6   X. Liu, R. K. Raj, T. J. Reichlmayr, C. Liu, and A. Pantaleev, "Teaching service-oriented programming to CS and SE undergraduate students," in *Frontiers in Education Conference, 2013 IEEE*, pp. 15–16.

7   X. Liu, R. Raj, T. Reichlmayr, C. Liu, and A. Pantaleev, "Incorporating service-oriented programming techniques into undergraduate CS and SE curricula," in *Frontiers in Education Conference, 2013 IEEE*, pp. 1369–1371.

8   J. Schneider, L. Greenwood, J. Rosenbeck, and M. Valentine. "Standards-based curriculum," Collaboratory for Resiliency and Recovery, Rochester Institute of Technology. https://www.rit.edu/engineeringtechnology/crr/grants-research/standards-based -curriculum (accessed Aug. 1, 2021).

9   K. A. Smith, S. D. Sheppard, D. W. Johnson, and R. T. Johnson, "Pedagogies of engagement: Classroom-based practices," *Journal of Engineering Education,* 94(1), pp. 87–101, 2005.

CASE STUDY #14

Computer Graphics Technology

Rosemary Astheimer, Purdue University

SUMMARY

We are in the midst of a Fourth Industrial Revolution (Industry 4.0), where the manufacturing industry is leveraging a digital product definition, referred to as Model-Based Definition (MBD), transforming traditional manual industry practices into automated tasks through machine-to-machine communication. The seamless transfer of information to enterprise stakeholders during all stages of a product's life cycle are captured in 3D computer-aided design (CAD) models. The CAD model acts as the authoritative source of information, removing the risk of error-prone data recreation, traditionally captured in a 2D drawing. The elimination of resulting disconnects that result from the existence of multiple derivative data sets is essential for the manufacturing industry to remain competitive.

Industry standards are vital to providing the language for capturing MBD information uniformly, as well as ensuring the data can be exchanged between disparate systems.

TARGET AUDIENCE

Engineering and technology majors interested in working in industry as a designer, machinist, metrologist, technical trainer, systems integration specialist, or applications engineer are the target audience. Students enrolled in MFET 20301 (previously CGT 20301), "Model-Based Definition," at the Purdue Polytechnic Institute have been learning about such practices since 2019.

INFORMATION LITERACY LEARNING OUTCOMES

The first learning outcome includes bringing awareness of the pervasiveness of standards that are used in nearly every aspect of our daily lives, from the design of a Universal Serial Bus (USB) device to certifications such as ISO 9001, which ensures consistent practices within a company.

At the center of MBD is the ASME Y14.5 standard, which defines symbology to capture product tolerance requirements at the engineering design level, referred to as product and manufacturing information (PMI). This PMI conveys required tolerances for specific features of a product to ensure it will perform its function as designed, which can then be used by manufacturing equipment, such as a computer numerical control (CNC) machine, to manufacture the product with the required precision. Inspection and quality control use this same information to drive automated inspection with a coordinate-measuring machine (CMM) to verify that the manufactured product meets the specifications that were set in the CAD model definition. The technical documentation that reports the results of an inspection are driven by the AS9102 standard.

The exchange of data between the stages of a product's life is typically not a direct translation, but rather there is a mechanism in place to translate the data. For example, the first word processing documents written on an Apple computer could not be opened on computers running the Windows operating system. Today, we take for granted that we have the ability to do a "Save As" on a text document and turn it into a web page, for example. Neutral file formats, such as STEP AP242 (ISO 10303-242), are vital to allow data transfer of CAD model information to be reused by the many systems involved.

Because significant change such as MBD takes time for industry to adopt, students also need to understand the history of product design as legacy data will be around for some time to come. Aircraft that are in service for decades are a prime example of a product that was produced before MBD, and the time to recreate such data in a MBD format is a non-value-added activity. After students learn what technical information needs to be captured to produce a product, the traditional practices that were used to capture that information—and how those practices were insufficient—must also be understood.

DESCRIPTION OF INSTRUCTION

Approximately 30 minutes of asynchronous lectures and videos of industry activists introduce a topic before each class. Articles, press releases, and use cases are assigned to demonstrate the effectiveness and use of industry design information. Students bring questions to class to discuss what they learned and what they found interesting. The instructor leads the discussion during one 50-minute class each week to aid with the understanding of the materials and demonstrates where appropriate.

Industry software vendors are invited to attend lectures to demonstrate and discuss the criticality, applicability, and conformance to standards of the data to the product design process. Publicly available online videos from vendors and users of large industrial equipment are integrated into the curriculum to demonstrate tools that cannot be brought into the classroom. Knowledge is applied through practice-based learning exercises for approximately two hours each week.

Students learn how to locate applicable standards documents through the Purdue University library and then follow those guidelines to apply product requirements in a CAD model. Industry software applications are used during the process to give students real-world, hands-on experience, including optimizing geometric design and verifying the data integrity of their solutions for conformance with standards. In the second half of the semester, previous exercise datasets are reused to generate programs to automate manufacturing, inspection, and generation of technical documents such as assembly instructions.

ASSESSMENT

As a final project, students write about a particular product, walking through its life cycle, presenting the information that is necessary at each stage, how that data is reused, and the importance and role applicable standards play in the success of these activities.

Students often contact me after they have graduated and entered the workforce to tell me how they have encountered and used standards that were first introduced in my class. This feedback demonstrates the value of practice-based learning and how it prepares students to hit the ground running, making them valuable assets and champions immediately for standardized methods in the workforce.

A study with International TechneGroup Incorporated (ITI) and the Department of Navy, Naval Air Warfare Center Aircraft Division (NAWCAD) [1] has shown that MBD results in significant savings of more than $3 million annually. Graduates who understand the concepts and the benefits of MBD will become advocates for this modernized method upon entry into the workforce and continue to drive efforts to advance it.

REFERENCE

1   ITI, Naval Air Systems Command (NAVAIR), Anark Corporation, The National Center for Manufacturing Sciences (NCMS), "US Navy model based definition initiative identifies significant savings," Milford, OH, 2013.

Acknowledgments

We would like to acknowledge the many people who made this book possible. First, thank-you to Dr. Clarence Maybee, editor of the Purdue Information Literacy Handbooks series, for your direction and encouragement throughout this process. For the opinions and ideas you shared from proposal submission all the way to book title finalization, we are incredibly grateful.

Our book was reviewed by many additional sets of eyes along the way, and we would also like to thank those who took the time to review the draft in full. Thank you to Karen Reczek, Standards Coordination Office, National Institute of Standards and Technology (NIST), who provided excellent insight and guidance based on her extensive standards background. You brought a wonderful perspective to what was included in the book, and we are grateful for your wisdom. Thank you also to Jill Hackenberg, computer science librarian at the University at Buffalo, for your copy editing. Your keen eye was a huge asset. We would also like to thank the anonymous peer reviewers for their time to read the entire draft and for providing many thoughtful suggestions. Your feedback was crucial to the publication of this book.

Many others assisted in the review of specific sections of the draft or provided other aid to the book. For their time, efforts, and contributions, we thank: Erin Smith, research and engagement librarian, Case Western Reserve University; Electra Enslow, director, Clinical Research and Data Services, University of Washington; Alex Carroll, librarian for STEM research, Vanderbilt University; Leena Lalwani, engineering librarian at University of Michigan; Paul Grochowski, engineering librarian at University of Michigan; and Emily Sartorius, graduate student at University of Michigan.

Index

Page numbers in italics indicate Figures and Tables.

search: ASSIST Quick Search and military standards, 47; engines, 53–54; for health care standards, 158; for standards, 57, 58–59. *See also* databases
Securities and Exchange Commission (SEC), 123
security, 22–23, 97, 113–15, 119, 124, 125
Security Standards Council Payment Card Industry (PCI) Data Security Standard, 124
Sentinel Initiative, FDA, 151, 158
SEP (standard-essential patent), 139
SES (Society for Standards Professionals), 96, 97
SI (International System of Units), 10
SIBR (Standards Incorporated by Reference) database, NIST, 48, 135
SNOMED-CT (Systematized Nomenclature of Medicine—Clinical Terms), 155, 156
social media feeds, 36
societal security, 115
Society for Standards Professionals (SES), 96, 97
Society of Automotive Engineering (SAE), 28, 36
software security, 114
software standards, 22, 110, 111
South Asian Regional Standards Organization (SARSO), 118
South Carolina Department of Transportation (SCDOT), 195–96
South Korea, 31, 33, 36
specifications, ICT standards, 111
Spectrum Quality Standards, 126
SpringShare LTI, 171
SSCD (Standards Services Curricula Development) Cooperative Agreement Program, NIST, 96
stakeholder benefits, standards, 11–12
standard-essential patent (SEP), 139
standardization, 3–8, 11–13, 31, 95, 117–18, 123. *See also* European Standardization

Organizations; International Organization for Standardization
Standardization Administration of the People's Republic of China (SAC, 2035 Standards Plan), 32
standardization development process: conformity assessment and certification, 35; dissemination, 35–36; government standards, 34; open standards, 34–35; of other standards, 30–31; SDOs and, 27–31, 35–36; in U.S. compared to other countries, 31–33; visualization, 29; voluntary consensus, 28–30
Standardization News magazine, 35
standardization process, 6, 16, 44–45, 96, 110–11, 178, 203, 211
Standard Practice for Thermal Diffusivity, Flash Method, 211, 212
standards, 9–11, 18; defined, 3, 14, 15–16, 53, 111–12; dissemination of, 28, 35–36, 214; by economic sector, 19–24; history, 3, 4–8, 12–13, 145–47; mandatory, 16–17, 30–31, 125; voluntary, 7, 16–17, 21, 110, 125, 128, 129. *See also specific topics*
standards, discovering and accessing: aggregator databases, 54, 55, *55*, 59; challenges, 57–60; costs, 57, 58; DRM restrictions, 60, 64; freely available full-text standards, 60, *61–63*, 63; full-text access restrictions, 57, 60; information literacy and access to, 48; library catalogs/discovery layers, 54, 55; library databases, 55, *57*; SDOs databases, 54, 55, 59; search engines, 53–54; searching for, 57, 58–59
Standards are Everywhere Tutorials, Purdue University, 97
Standards Aware, 97
standards collection development: aggregator databases and services, 70–71; IHS Global standards store, 71–72; models of, 68–72; PDFs purchased individually and electronically, 69, 70;

About the Editors

Chelsea Leachman is a science and engineering librarian at Washington State University. She liaises with biological systems engineering, civil and environmental engineering, computer science, electrical engineering, mathematics, mechanical engineering, and materials engineering.

Erin M. Rowley is the head of science and engineering library services and is the engineering librarian at the University at Buffalo, working with students, faculty, and staff in the School of Engineering and Applied Sciences. Before joining the university, she worked as a corporate research librarian at a consumer products testing laboratory for nearly nine years using standards from around the world on a daily basis.

Margaret Phillips is an engineering information specialist and associate professor in the Purdue University Libraries and School of Information Studies. She is the liaison to the engineering technology, industrial engineering, and nuclear engineering departments, and acts as the standards librarian.

Daniela Solomon is a research and engagement librarian at Case Western Reserve University, where she is the liaison to the Case School of Engineering, and manages the on-demand standards service.